EXTENDED HEREDITY

EXTENDED HEREDITY

A NEW UNDERSTANDING OF INHERITANCE AND EVOLUTION

Russell Bonduriansky and *Troy Day*

PRINCETON UNIVERSITY PRESS
PRINCETON AND OXFORD

Published by Princeton University Press,
41 William Street, Princeton, New Jersey 08540

In the United Kingdom: Princeton University Press,
6 Oxford Street, Woodstock, Oxfordshire OX20 1TR

press.princeton.edu

Jacket images courtesy of Shutterstock

Library of Congress Control Number 2017958983
ISBN 978-0-691-15767-2

British Library Cataloging-in-Publication Data is available

This book has been composed in Minion Pro

Printed on acid-free paper. ∞

Printed in the United States of America

10 9 8 7 6 5 4 3 2 1

To our children, Aaron and Amalyn, Willem and Samantha,
who are more than the sum of their genes

CONTENTS

PREFACE

The nature of heredity—that is, how biological information is transmitted across generations—is a question that touches just about every part of the biomedical sciences, from evolutionary biology's quest to account for the diversity of life to medical science's effort to understand why certain diseases run in families. It's also widely seen as an iconic success story of modern science. From the cautious speculations of the nineteenth century to the establishment of Mendelian genetics in the early twentieth century and the deciphering of the genetic code in the 1960s, the unlocking of "the mechanism of heredity" is portrayed in textbooks as a journey now essentially completed.

But nature often manages to frustrate our desire for simple answers. Several decades of troubling discoveries that don't fit the established picture of how the world is supposed to work are now leading some scientists to argue that it's time to rethink the nature of heredity. If this challenge succeeds, then the biological and medical sciences will be in for a shake-up over the coming years, and even the textbooks will have to be rewritten.

If we were to try to summarize the main thesis of this book in a few words it would be like this: there is more to heredity than DNA sequences (genes), and recognizing this nongenetic dimension of heredity can provide us with new insights into how evolution works, and into many practical concerns of human life as well. It's now clear that a variety of nongenetic factors are transmitted across generations alongside genes. Like genes, such nongenetic factors can convey biological information across generations, confer a resemblance between offspring and their parents, and potentially influence the course of evolution. This plurality of hereditary factors is necessitated by basic properties shared by all cellular life-forms, and we believe that any concept of heredity that is not arbitrarily narrow must encompass this nongenetic dimension. A concept of heredity that encompasses both genetic and nongenetic processes is already emerging, and we will refer to it as "extended heredity" to differentiate it from the conventional, genocentric view.[1]

The narrow concept of heredity that has prevailed since the early decades of the twentieth century has resulted in the exclusion from the realm of possibility of some very real and important biological phenomena, such as the possibility that an individual's experiences during its lifetime can have predictable consequences for the features of its descendants. Such effects were long dismissed as a "chemical impossibility" and a violation of the "central dogma" of molecular biology. Yet, a great variety of such phenomena have now been reported in the scientific literature. This nongenetic side of heredity has been a blind spot for biology and medicine for decades, but the elephant in the room is finally starting to be noticed. For readers who are new to this field, we aim to provide a way to think about these recent developments, place them in historical context, and understand their implications. For those who remain skeptical of such heterodox ideas, we hope to at least bring the issues into sharper focus.

We are not the first to recognize the implications of nongenetic inheritance. In this book, we draw on numerous lines of research by many authors, borrowing from such far-flung areas as cultural inheritance theory, niche construction theory, evolutionary ecology, and molecular and cell biology. Work in all these areas has addressed various aspects of what the classic narrative leaves out.[2] Most importantly, two thought-provoking books by Eva Jablonka and Marion Lamb—*Epigenetic Inheritance and Evolution* (1995) and *Evolution in Four Dimensions* (2005)—have focused squarely on the implications of nongenetic inheritance for evolution. Jablonka and Lamb's books and papers paved the way for the recent upsurge in evolutionary research on extended heredity, and crystallized many of the ideas explored by subsequent authors, including ourselves. Yet, while we acknowledge our intellectual debt to Jablonka and Lamb, this book reflects our own perspective, approach, and aims. In particular, building on the conceptual groundwork laid by previous authors, our main objective is to explore how an extended concept of heredity can be incorporated into evolutionary theory and why doing so can provide new insight on a range of evolutionary questions.

We have subdivided this book into ten chapters.

In chapters 1 ("How to Construct an Organism") and 2 ("Heredity from First Principles"), we explain why the classic framework is overly narrow, and why we believe that an extended concept of heredity is necessitated by universal features of cellular life. These chapters introduce ideas that are developed more fully in the rest of the book.

In chapter 3 ("The Triumph of the Gene"), we explore the development of the modern, genocentric concept of heredity. Although scientists rarely bother to delve into the historical baggage of their discipline, we believe that without understanding the history it is all but impossible to understand current developments in this field. For example, why is it that nongenetic inheritance was so unequivocally rejected by leading twentieth-century biologists? And why are their arguments no longer valid today?

Chapters 4 and 5 provide an overview of the evidence for nongenetic inheritance and illustrate its diversity and importance. Chapter 4 ("Monsters, Worms, and Rats") focuses on fascinating discoveries of a type of nongenetic inheritance, known as "epigenetic inheritance," that has recently been receiving a lot of attention in medical science as well as the mainstream media. In chapter 5 ("The Nongenetic Inheritance Spectrum"), we show that phenomena that are "epigenetic" in the strict sense are part of a much broader array of nongenetic inheritance mechanisms that are just as interesting and important.

The next set of chapters explores the implications of extended heredity in more depth. In chapter 6 ("Evolution with Extended Heredity"), we show how the ideas developed in earlier chapters can be incorporated into evolutionary theory, furnishing a framework that allows us to explore their consequences for evolution. In chapter 7 ("Why Extended Heredity Matters"), we use this framework to illustrate how nongenetic inheritance can change the trajectory and outcome of evolution. In chapter 8 ("Apples and Oranges?"), we confront the key criticisms that evolutionary biologists have leveled against extended heredity. In chapter 9 ("A New Perspective on Old Questions"), we revisit some of the thorniest puzzles in evolutionary biology and show how the insights provided by extended heredity can allow us to see these questions in a new light.

Finally, in chapter 10 ("Extended Heredity in Human Life"), we consider the implications of extended heredity for the practical concerns of modern human beings living in a rapidly changing world. We show that, during its heyday in the twentieth century, the narrow, genocentric view of heredity had very tangible and sometimes tragic consequences, and we consider how extended heredity might alter our understanding of our health and society, and our impact on the world around us.

We have endeavored to make this book as accessible as possible in the hope that it will be read not only by practicing biologists but also by students and laypeople who follow biology. While jargon is sometimes unavoidable, we have made an effort to define all technical terms (with some definitions and explanations provided in notes at the back of the book). Where mathematical ideas form an essential part of the story, we have tried to present them in intuitive and pictorial ways. Equations have been kept to a minimum and are mostly consigned to boxes and notes that lay readers can safely skip without losing the bigger plot.

We devote relatively little space to discussion of molecular mechanisms. Our main aim in this book is to explore the implications of extended heredity for evolution and, for this reason, we focus on effects at whole-organism and ecological levels and provide just enough detail on proximate mechanism to allow readers to understand the general nature of these effects. Moreover, molecular biology is developing at such a breathtaking pace that any details we provide are likely to be outdated by the time this book rolls off the printing press. Readers who wish to delve deeper into the details of molecular mechanism can easily find up-to-date reviews.

We should also state at the outset that the ideas presented in this book do not refute evolutionary theory or the central role that genetics plays in it. We see genetic and nongenetic inheritance as hereditary processes that operate in parallel, so extended heredity supplements rather than supplants genetics. Likewise, although we believe that these ideas have important implications for evolutionary biology, extended heredity does not challenge Darwin's basic insight that natural selection coupled with inheritance is the primary cause of adaptive evolution.

Who are we? RB is an evolutionary biologist in the Evolution and Ecology Research Centre and School of Biological, Earth, and Environmental Sciences at the University of New South Wales in Sydney,

Australia. TD is cross-appointed between the Departments of Biology and Mathematics at Queen's University in Kingston, Canada. Since meeting at the University of Toronto around the turn of the century, we have collaborated on a number of research projects and, somewhere along the way, became interested in extended heredity and its implications for evolution. This book is the culmination of several years of collaborative research on this problem.

EXTENDED HEREDITY

1

How to Construct an Organism

What I cannot create I do not understand.
—Richard P. Feynman[3]

Not so long ago, newspaper headlines around the world proclaimed that scientists had created "artificial life." This astonishing news referred to an experiment from the laboratory of maverick molecular biologist Craig Venter, in which the DNA molecule of a simple type of bacteria had been artificially synthesized from its chemical building blocks (with some curious embellishments, like Venter's email address encrypted in the DNA's genetic code), and then inserted into a different species of bacteria, replacing that cell's own genome. Amazingly, this procedure resulted in a living bacterial cell that went on to divide and produce a colony of bacteria.[4]

Beyond its sheer technical wizardry, Venter's experiment seems to offer a unique insight into the nature of heredity—the transmission of biological information across generations that causes offspring to resemble their parents, and can thereby enable evolution by natural selection.[5] After all, Venter's research group had managed to decouple two fundamental components of a cellular organism—the genome (that is, the DNA sequence) and its cytoplasmic surroundings (that is, the immensely complex biomolecular machinery that constitutes a living cell). The resulting bacterial chimera, which combines the genome of one species with the cytoplasm of another, should therefore tell us something about the roles of the DNA sequence and the cytoplasm in the transmission of organismal traits across generations. Did Venter's bacterium resemble the species from which it got its DNA sequence, the species from which it got its cytoplasm, or both?

Reports on Venter's experiment emphasized the role of the genome in converting the bacterial host cell into a different species of bacteria: the genome induced changes in the features of the cell into which it had been inserted, such that, after several cycles of cell division, the descendants of the original chimeric cell came to resemble the genome-donor species. This result illustrates the DNA's well-known role in heredity: the base-pair sequence of the DNA molecule *encodes* information that is *expressed* in the features of the organism. Indeed, from here, it seems a small step to conclude that the cytoplasm (and, by extension, any multicelled body) is fully determined by the genome, and that the DNA sequence is all we need to know to understand heredity. Venter's experiment thus seems to provide a powerful confirmation of a concept of heredity that has underpinned genetics and evolutionary biology for nearly a century.

But take a closer look at Venter's experiment and the picture becomes less clear. Although many media reports gave the impression that Venter's "artificial" organism was created from a genome in a petri dish, the bacterial chimera actually consisted of a completely natural bacterial cell in which only one of many molecular components had been replaced with an artificial substitute. This is an important reality check: although it's now possible to synthesize a DNA strand, the possibility of creating a fully synthetic cell remains the stuff of science fiction.[6] In fact, rather than demonstrating the creation of artificial life, Venter's experiment neatly illustrates a universal property of cellular life-forms: all living cells come from preexisting cells, forming an unbroken cytoplasmic lineage stretching back to the origin of cellular life. This continuity of the cytoplasm is as universal and fundamental a feature of cellular life-forms as the continuity of the genome. Of course, cytoplasmic continuity does not in itself demonstrate that the cytoplasm plays an independent role in heredity. After all, the features of the cytoplasm could be fully encoded in the genes. Yet, the potential for a nongenetic dimension of heredity clearly exists.[7]

The continuity of the cell lineage has been recognized since the mid-nineteenth century but, since the dawn of classical genetics in the early twentieth century, many biologists have been at pains to deny or downplay the role of nongenetic factors in heredity, arguing that the transmission of organismal features across generations results more or less entirely from the transmission of genes in the cell nucleus.[8] Genes were assumed to be impervious to environmental influence, so that an individual could only transmit traits that it had itself inherited from its parents. These ideas gained prominence while the term "gene" still referred to an entirely theoretical entity, and long before molecular biologists uncovered DNA's structure and the genetic code. More recently, this view was popularized by Richard Dawkins in his memorable image of the body as a lumbering robot built by genes to promote their own replication. But this purely genetic concept of heredity was never firmly backed by evidence or logic. Venter's chimeric bacteria were foreshadowed by late nineteenth-century embryological experiments that combined the cytoplasm of one species with a nucleus from another species, providing the first hints that the cytoplasm is not a homogeneous jelly but a complex machine whose components and three-dimensional structure control early development. Further tantalizing hints of a nongenetic dimension to heredity were provided by the work of mid-twentieth-century biologists who discovered that mechanical manipulation of the structure of single-celled organisms like *Paramecium* could result in variations that were passed down unchanged over many generations. Today, after many more clues have come to light, biologists are finally beginning to reconsider the possibility that there is more to heredity than genes.

RETURN OF THE NEANDERTHALS?

Venter's experiment raises intriguing questions about the nature of heredity at the level of a single cell, but what about multicelled organisms like plants and animals? A single cell's cytoplasm is divided in half each time the cell divides and then supplemented with newly synthesized proteins encoded by the genome. It is this process of gradual conversion that allowed the bacterial genome to gradually reset features of the host cell in Venter's experiment. Can such conversion also reset the features of more complex life-forms?

Consider an example at the opposite extreme of the complexity gradient—the recent idea of resurrecting a Neanderthal. Some people believe that such a feat could be accomplished by implanting a synthetic

Neanderthal genome (whose sequence was recently deciphered from DNA fragments extracted from ancient bones) into a modern human egg or stem cell deprived of its own genome. Ethical considerations aside, it would be extremely interesting to compare the physical and mental traits of our enigmatic sister species with our own, and on the face of it, such an experiment could be carried out by following Venter's recipe. What's less clear is how closely the resulting creature would resemble a genuine Neanderthal.

Neanderthals differed from us *Homo sapiens* in many features of their bodies, such as their muscular build, long, low skulls with heavy brow ridges, and more rapid juvenile development[9] (figure 1.1). Some paleoanthropologists also believe that Neanderthals differed from contemporaneous *Homo sapiens* populations in various aspects of their culture and social organization, such as their use of clothing, foraging techniques, and reliance on long-distance trading networks.[10] Which of these features could we expect to observe in an individual derived from a Neanderthal genome implanted into a modern human egg?

Clearly, such a creature would fail to exhibit Neanderthal cultural practices, since culture is not encoded in the genes (although a population of such creatures, if allowed to interbreed for many generations in isolation, could perhaps tell us something about Neanderthals' capacity to develop complex culture). A lone Neanderthal growing up playing video games and watching movies in its enclosure at the primate research institute would surely fail to develop many of the behavioral peculiarities of its species. Moreover, we know that physical activity influences the development of bones and muscles, while dietary preferences and practices (which are partly culturally transmitted) influence the development of dental and cranial features. So even the distinctive features of Neanderthal bodies may have been a product not only of Neanderthal genes but also of how they behaved and what they ate. A couch-potato Neanderthal will undoubtedly exhibit some of the distinctive features of Neanderthal physiology but might still end up looking more like a specimen of modern, industrialized *Homo sapiens*, with its proverbial joy-stick thumb, fondness for potato chips, and alarming body-mass index.

But the problem runs even deeper. In all complex organisms, development is largely regulated by *epigenetic* factors—molecules (such as methyl groups and noncoding RNAs) that interact with the DNA and

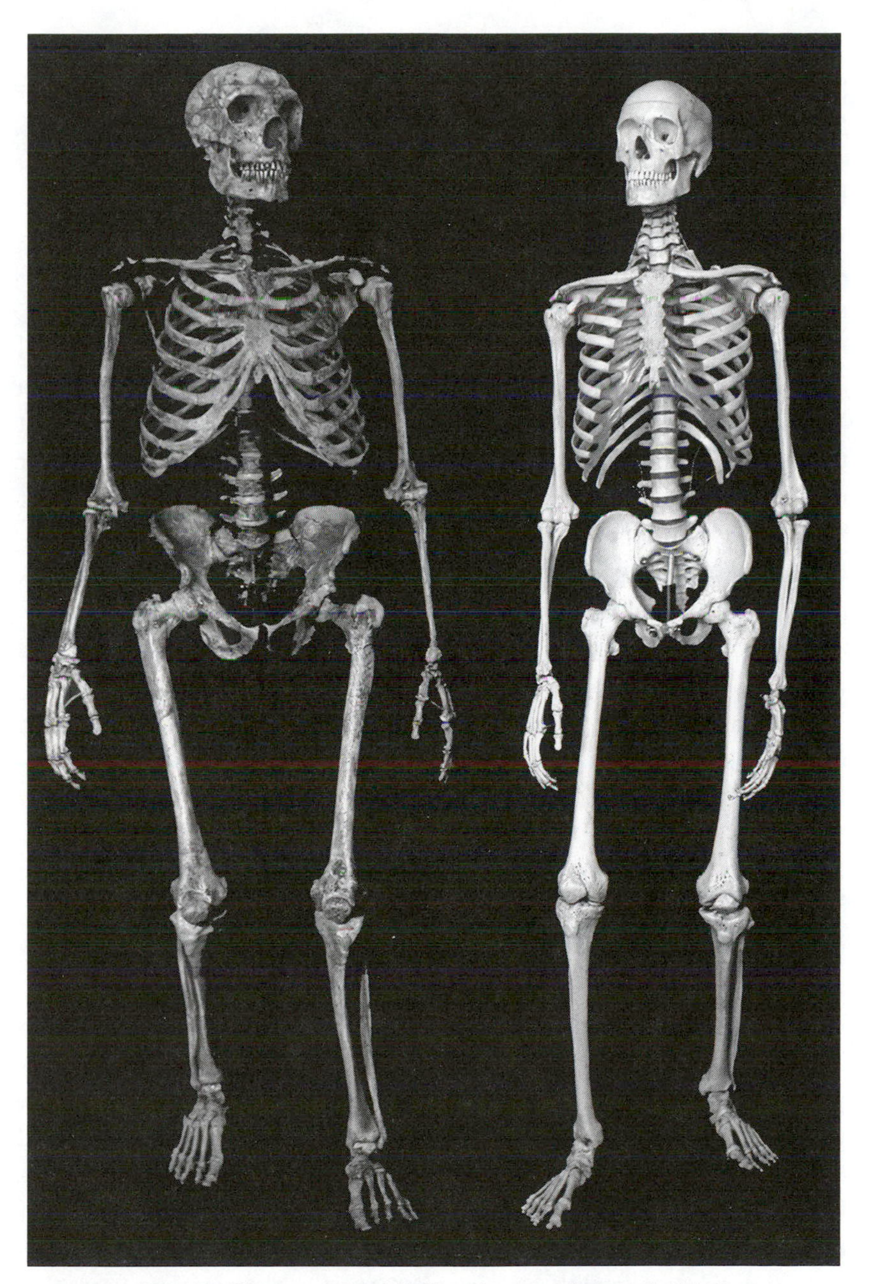

Figure 1.1. Skeletons of a Neanderthal (*left*) and modern human (*right*). Can a Neanderthal be resurrected by implanting a Neanderthal DNA sequence into a modern human egg? (© I. Tattersall, Photo: K. Mowbray)

influence when, where, and how much genes are expressed. Some epigenetic factors can be acquired through exposure to particular environmental factors such as diet, and can then be transmitted to offspring. Although recent research by Liran Carmel's lab in Israel has begun to uncover aspects of the Neanderthal epigenome,[11] it remains unclear which differences between Neanderthals and *Homo sapiens* were downstream consequences of genetic differences and which differences resulted from their long-vanished environment and lifestyle. Indeed, some epigenetic patterns found in children conceived during seasonal cycles of food shortage in an agricultural population in The Gambia in West Africa were also characteristic of Neanderthals, suggesting that these epigenetic features of Neanderthals may have been a result of their diet rather than their genes.[12] Unless such epigenetic factors, and other nongenetic influences on development such as cytoplasmic and intrauterine factors, can be reconstructed along with the Neanderthal DNA sequence, our Neanderthal may lose even more of its distinctive traits.

In short, we suspect that implanting a Neanderthal genome into a modern human egg would result in a creature that diverged in many behavioral and physical features from genuine Neanderthals. The reason for this is simply that a DNA sequence does not contain all the information needed to re-create an organism.

WHY NOTHING IN BIOLOGY MAKES SENSE ANYMORE

The idea that genes encode all the heritable features of living things has been a fundamental tenet of genetics and evolutionary biology for many years, but this assumption has always coexisted uncomfortably with the messy findings of empirical research. The complications have multiplied exponentially in recent years under the weight of new discoveries.

Classical genetics draws a fundamental distinction between the "genotype" (that is, the set of genes that an individual carries and can pass on to its descendants) and the "phenotype" (that is, the transient body that bears the stamp of the environments and experiences that it has encountered but whose features cannot be transmitted to offspring). Only those traits that are genetically determined are assumed to be heritable—that is, capable of being transmitted to offspring—because inheritance occurs exclusively through the transmission of genes. Yet, in violation of the genotype/phenotype dichotomy, lines of genetically identical animals and plants have been shown to harbor heritable variation and respond to natural selection. Conversely, genes currently fail to account for resemblance among relatives in some complex traits and diseases—a problem dubbed the "missing heritability."[13] But, while an individual's own genotype doesn't seem to account for some of its features, parental genes have been found to affect traits in offspring that don't inherit those genes. Moreover, studies on plants, insects, rodents, and other organisms show that an individual's environment and experiences during its lifetime—diet, temperature, parasites, social interactions—can influence the features of its descendants, and research on our own species suggests that we are no different in this respect. Some of these findings clearly fit the definition of "inheritance of acquired traits"—a phenomenon that, according to a famous analogy from before the Google era, is as implausible as a telegram sent from Beijing in Chinese arriving in London already translated into English.[14] But today such phenomena are regularly reported in scientific journals. And just as the Internet and instant translation have revolutionized communication, discoveries in molecular biology are upending notions about what can and cannot be transmitted across generations.

Biologists are now faced with the monumental challenge of making sense of a rapidly growing menagerie of discoveries that violate deeply ingrained ideas. One can get a sense of the growing dissonance between theory and evidence by perusing a recent review of such studies and then reading the introductory chapter from any undergraduate biology textbook. Something is clearly missing from the conventional concept of heredity, which asserts that inheritance is mediated exclusively by genes and denies the possibility that some effects of environment and experience can be transmitted to descendants.

In the following chapters, we will sketch the outlines of an extended concept of heredity that encompasses both genetic and nongenetic factors and explore its implications for evolutionary biology and for human life.

2

Heredity from First Principles

The whole subject of inheritance is wonderful.
—Charles Darwin, *Variation of Animals and Plants under Domestication*, 1875

If there is one property that captures the uniqueness of living things, it is their ability to perpetuate their kind through the production of similar forms—that is, reproduction with heredity.[15] In all cellular life-forms (that is, all but the simplest biological entities, such as viruses), biological reproduction also follows a universal pattern that can be said to comprise two basic elements. First, reproduction involves the perpetuation of the cell lineage through an unbroken chain of cell division, such that all cells (including Venter's chimeric bacteria) come from preexisting cells.[16] Second, reproduction involves the duplication and transmission of a DNA sequence, embodied in the famous double helix whose chemical properties encode "instructions" for the synthesis of proteins and the regulation of cellular processes. To us, these two basic elements of the reproductive process imply an inherent duality in the nature of heredity (figure 2.1).

In this chapter, we will attempt to reimagine heredity from first principles. The point of this somewhat quixotic exercise is to walk the reader through the logic of extended heredity and (we hope) make a convincing case for the ideas that we will elaborate upon and apply later in this book. These ideas are not really new. Although, as we will see in chapter 3, the triumph of Mendelian genetics in the early twentieth century displaced the debate on the nature of heredity to the margins of biology, calls to extend heredity to encompass nongenetic factors

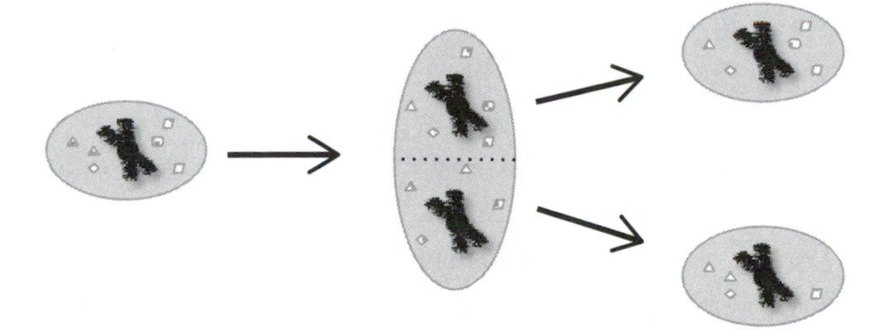

Figure 2.1. A schematic of the duality of heredity in its simplest form. DNA sequences are represented by the black chromosome, and nongenetic material is represented by the small triangles, diamonds, and other shapes in the cell's cytoplasm. In sexually reproducing organisms, a variety of nongenetic factors are transmitted inside or along with the gametes. In many multicellular organisms, a diverse array of nongenetic factors can also be transmitted through postfertilization parent-offspring interactions and parental investment. All of these processes can mediate the transmission of variation across generations and can therefore be viewed as mechanisms of inheritance. We refer to the totality of these genetic and nongenetic mechanisms of inheritance as "extended heredity."

alongside genes continued into the 1960s.[17] This debate resumed in the 1990s as evidence of inheritance through epigenetic mechanisms such as the transmission of DNA methylation patterns (that is, the presence or absence of methyl groups bonded to certain DNA bases) began to emerge.[18] Yet, scientists being a cautious and conservative tribe, the idea of extended heredity is only now starting to be taken seriously, and the outlines of this new concept are still very much in flux.[19]

We will begin by examining the genetic and nongenetic components of heredity and then consider how these components can be combined into a concept of extended heredity.

THE GENETIC LIBRARY

DNA is a critically important component of the cellular machinery that regulates the physiological processes and responses that take place within a single cell or a multicelled organism, from juvenile development and growth, to reproduction and, ultimately, aging and death. The genetic information encoded in the base-pair sequence of the DNA—that is, the genome and its constituent genes—serves as a molecular library in which the amino-acid sequences of all bodily proteins, as well as "noncoding" regulatory instructions, are stored in a "genetic code." (Many genomes, including ours, also contain large quantities of "junk DNA" that does not seem to serve any function for the organism, including parasitic DNA sequences called "transposable elements" that can insert new copies of themselves within the genome.) A complex molecular machinery first "transcribes" a DNA sequence into a corresponding RNA strand, and then "translates" the sequences of RNA bases into a sequence of amino acids that ultimately form a protein. This process, called "gene expression," is exquisitely sensitive to both internal state (for example, health, hunger, age) and to input from the external environment. For example, eating foods rich in protein stimulates the expression of the *IGF1* gene in the liver, causing the liver to secrete a protein called insulin-like growth factor 1, an important hormone that stimulates growth in childhood but that can also promote aging in adults. However, the nature and strength of this response also depends on the base-pair sequences of other genes, such as the gene encoding the growth hormone receptor.[20]

DNA plays a key role in heredity as well. Individuals vary in the DNA sequences within their genomes and can transmit these variable sequences (called "genetic alleles") to their offspring. When a cell divides to give rise to two daughter cells, both cells receive copies of the DNA from the original cell. When egg and sperm fuse, the newly formed offspring receives partial, complementary copies of the genomes of its parents. We resemble our parents partly because we carry genetic alleles that we inherited from them.

To appreciate the unique role of the genes, it's important to note three crucial attributes of DNA. First, the DNA molecule is remarkably stable—so much so that a lively research field has grown up around the study of ancient DNA fragments that can be extracted from bones or soft tissues of extinct species like Neanderthals and mammoths.[21] This chemical stability allows DNA to serve as a dependable library of information within the cell. Second, the DNA's double-stranded structure, in which each strand serves as a template for its complementary strand, enables the DNA to be replicated with very high fidelity, allowing genes to be passed down unaltered from parents to their offspring.[22] Random changes in the DNA sequence (mutations) occur throughout most of the human genome at a rate of less than one mutation for every ten million base pairs per generation.[23] The DNA's replication fidelity is so great that relatives of many human genes can be identified in yeast, showing that some DNA sequences from the single-celled common ancestor of humans and yeast have been transmitted across generations with little change for well over a billion years.[24] Third, DNA can store vast amounts of information. DNA is a molecular chain made up of just four types of nucleotide bases (the familiar A, T, G, and C, which stand for the chemical compounds adenine, thymine, guanine, and cytosine), but the chain can be extremely long (one set of human chromosomes, unraveled and strung end to end, would be over 1 meter in length), and can therefore encode a massive amount of information in the sequence of nucleotides. The number of possible ways to order the three billion base pairs contained in the human genome (that is, its "combinatorial complexity") is unimaginably vast. DNA's chemical stability and information storage capacity are so impressive that biotechnologists are even exploring the possibility of using DNA as a medium for data storage.[25]

Philosopher Kim Sterelny and colleagues have argued that the DNA's unique features endow it with a special evolved role in heredity, pointing out that life could not exist without a DNA-like system for encoding organismal features and enabling the transmission of these features across generations.[26] Within a living body, the maintenance of the intricate and fragile systems that enable survival and reproduction result from a continued balance between the degrading effects of mutation and the restorative effects of cellular mechanisms that repair damaged DNA or remove damaged cells. Within a population, an analogous balance must exist between the degrading effects of mutation and the purifying effects of natural selection, which removes individuals bearing deleterious mutations from the gene pool. Without a sufficiently high degree of stability and replication fidelity, the rate of mutation would outpace the effects of natural selection, and the intricate organization of living things would not be possible. Likewise, without sufficient combinatorial complexity, DNA would not be able to encode the features of life-forms as complex and different as slime molds and blue whales, or the extensive variation in genetically based features found within every biological population.

NONGENETIC FACTORS IN HEREDITY

DNA plays a central role in heredity and development, but is it the whole story? Popular science and the mainstream media often seem to endow DNA with almost magical qualities, claiming that DNA can "replicate itself" or that scientists have discovered a "gene for" intelligence, religiosity, political affiliation, or criminality.[27] Such claims embody the deep-seated belief that DNA is the essence of life, the sole determinant of organismal features, and the exclusive basis of heredity. These popular notions are oversimplifications, but they undoubtedly have their roots in the purely genetic concept of heredity that has dominated biology since the early twentieth century.

The reality is that, by itself, DNA can't replicate, can't make you smart, and can't endow your children with your intellectual prowess.[28] DNA is just one component in a complex, highly structured biochemical machine whose properties emerge not only from the nature of its parts but also from the precise juxtaposition and interactions of these parts. Fertilization involves the fusion of two cells that contain thousands of different kinds of biomolecules besides DNA. The genocentric view of heredity is thus based on the implicit assumption that the many other factors transmitted from parents to offspring can be safely ignored, presumably because they are mere downstream products of gene expression and that any variation in such nongenetic factors is therefore fully encoded in the genes. As we will see, this assumption was never well supported and is becoming ever more untenable under the weight of new evidence.

DNA is often regarded as a purely digital, linear medium of information storage whose information content is fully embodied in its base-pair sequence, in the same way that the information in this book is embodied in a sequence of letters. However, research over the past several decades has revealed that DNA invariably comes with an "epigenetic" overlay of molecules, such as methyl groups, histone proteins, and RNA, that affect how DNA sequences are "read" by the cell. In other words, epigenetic factors play an essential role in regulating gene expression. It is this epigenetic machinery that allows a single genome to produce a diverse array of cell and tissue types such as muscles, neurons, and blood during embryonic development, and also allows organisms to modulate gene expression in response to their surroundings and activities. Crucially, as we will see in chapter 4, accumulating evidence shows that epigenetic factors can change spontaneously or respond in consistent ways to the ambient environment, and, in some cases, such epigenetic variants can be transmitted across generations. In other words, epigenetic variation is partially independent of the genes and, when transmitted across generations, such variation can affect offspring development—a phenomenon known as "epigenetic inheritance."[29]

Although its discovery has generated much excitement, epigenetic inheritance is only the tip of the nongenetic iceberg. Like all cells, eggs and sperm contain a cytoplasm and cell membrane made up of many different biomolecules such as proteins and lipids, and complex subcellular structures such as ribosomes, centrioles, mitochondria, and chloroplasts. Some structural information, including the precise asymmetrical architecture of the microtubule "skeleton" that pervades the cytoplasm, the topography of the membrane that encloses the cell, and even the three-dimensional structure of certain proteins, can also be transmitted from mother to daughter cells independently of genes, and the structure of the fertilized egg plays an important role in early embryonic development.[30] Furthermore, while a male's role in conception has traditionally been seen as the provision of genes, recent evidence suggests that the molecular cargo of the seminal fluid—the liquid medium in which sperm cells are transported—can influence offspring development as well.[31]

But, as every human parent knows, the job does not end at conception. In many animals, offspring develop inside their mother, or are cared for after birth by one or both parents. Many substances—milk and other glandular secretions, hormones, nutrients, symbionts, and pathogens—are transferred from parents to offspring after conception. Parents can also influence the development of their offspring by selecting and shaping the ambient environment in which the offspring begin their lives and, in animals with complex nervous systems, parents can influence what their offspring learn. Thus, genes are transmitted alongside a complex array of epigenetic, cytoplasmic, somatic, behavioral, and environmental factors, many of which can influence offspring development. We will explore the nature of these factors and their effects in more detail in chapters 4 and 5.

Because, like genes, some nongenetic factors vary among individuals, can be transmitted across generations, and can affect the features of descendants, we and a number of other researchers believe that it's appropriate to include such nongenetic factors within the scope of *heritable variation*. The application of this term to nongenetic factors deviates from a convention that has prevailed for many decades, but we believe that this convention is due for a revision. Some may regard this as merely a semantic issue, but words matter. Nongenetic variation has long been *defined* as nonheritable, and the assumption that environmentally induced traits and other forms of nongenetic variation cannot be transmitted to descendants has had a profound influence on the biomedical sciences.

The nongenetic factors and processes outlined previously clearly encompass a diverse range of cellular, physiological, or behavioral processes. Some of these processes (such as those involving the transmission of cytoplasmic and epigenetic variation) arise from fundamental biological properties shared by all eukaryotes or even all cellular lifeforms, while other processes, such as learning, operate only in some groups of organisms.[32] Nonetheless, we believe that all such "mechanisms" of heredity also have a number of properties in common, such as relatively high mutability, and the potential to transmit traits induced by environment and experience across generations.[33] Therefore, at least for some purposes, we believe that it makes sense to treat such mechanisms as a set under the common rubric of "nongenetic inheritance."[34] In particular, as we will illustrate in this book, doing so can help us to investigate the roles that nongenetic inheritance can play in evolution.

How does nongenetic inheritance work, and how do its properties compare with those of the more familiar mechanism of genetic inheritance? Just as genetic variation arises through random changes in DNA sequence (mutation), nongenetic variation can arise spontaneously, through random changes in the epigenome or other nongenetic factors that can be transmitted to offspring. Moreover, as with genetic inheritance, what parents transmit nongenetically to their offspring is usually not a phenotypic feature itself, but the potential to develop a similar feature. For example, we would say that body size is inherited genetically if some parents transmit alleles to their offspring that cause rapid growth. Similarly, body size might be inherited nongenetically if large parents transmit abundant nutrients to their offspring, thereby inducing rapid growth that results in large body size. In both cases, parents transmit factors (be they genetic alleles or nutrients) that influence offspring development so as to produce a particular phenotype (large body size) in the offspring.[35]

However, unlike genetic mutations, many types of nongenetic changes can be predictably induced by exposure to a particular environment, experience, or simply the effects of aging, and can then be transmitted to offspring. Nongenetic inheritance thus allows for the "inheritance of acquired traits." Although biology textbooks ritualistically assert that inheritance of acquired traits is impossible, we will encounter many examples of such inheritance in chapters 4 and 5. The inheritance of acquired traits can be likened to phenotypic plasticity (the sensitivity of growth and development to environment) except that here the plastic response extends across one or more generations.[36] And just as with phenotypic plasticity, it is important to distinguish between a response that is adaptive and one that is not. As we will see, there is ample evidence for the inheritance of acquired traits that are random with respect to their fitness consequences, and natural selection can act on these just as it acts on random genetic mutations. However, in some cases, nongenetic transmission can play an adaptive role analogous to adaptive forms of within-generation plasticity. Such fitness-enhancing forms of nongenetic inheritance are called "adaptive parental effects."[37]

The environment can also induce more subtle effects that are not visible in the parents but become apparent in offspring.[38] In other words, environmental factors can endow individuals with a propensity to transmit traits that they themselves do not express. For example, while cigarette smoking is bad for both mothers and their children, the effects are quite different: a woman who smokes can develop a range of respiratory, circulatory, and other physiological ailments, whereas embryos exposed to nicotine in utero may undergo epigenetic "reprogramming" at certain genes, resulting in developmental outcomes such as reduced birth weight and increased risk of behavioral disorders.[39] Such effects can be likened to a genetic mutation in the germ line (the specialized tissue that produces eggs or sperm), which has no visible effect on the parent but that can affect the features of its offspring.[40]

Aside from the potential to transmit some effects of environment, experience, and age, nongenetic factors also generally differ from genes in being less stable (or more "mutable") across generations than genetic alleles. For example, in many mammals, acquired tastes for a particular type of food can be transmitted from mother to offspring, either through learning or via fetal exposure to food-derived chemicals that pass from maternal to fetal blood via the placenta.[41] But transmission is probably limited to a single generation: if the offspring itself fails to encounter the food often enough, it will not transmit the preference to its own young, and many acquired traits are therefore easily lost.

However, some acquired traits may persist for many generations because they are self-regenerating. How does self-regeneration occur? For DNA sequences, self-regeneration depends on a highly precise copying mechanism involving specialized enzymes, equally specialized repair mechanisms that take advantage of the presence of two complementary strands (redundancy), and the chemical stability of the DNA molecule itself. For many nongenetic mechanisms, the transmission process is even more complex. Many nongenetic factors are much less stable than DNA, so the ability of these factors to persist across multiple generations is less straightforward to explain. Moreover, while genetic inheritance actually involves the transmission of DNA from parents to their offspring, nongenetic transmission and self-regeneration are sometimes indirect, comprising causal pathways that involve both soma and germ line.

The most intuitive examples are self-regenerating behaviors. For example, a preference for a particular food learned from the mother or acquired in utero will tend to regenerate itself when the offspring seeks out that particular food. The offspring will then transmit the preference by similar mechanisms to its own offspring. In the realm of behavior or culture, such sequences of acquisition and transmission can be stable over multiple generations, as family traditions sometimes are in humans.[42] However, both the learning-mediated and intrauterine environment-mediated transmission of diet preference involve indirect mechanisms of self-regeneration, in that the transmission of the physiological or cognitive trait (the diet preference) depends on the successful execution of a behavior (locating and ingesting the food in question) on the part of the offspring. An individual that inherits a craving for a particular food but is unable to actually feed on it will fail to transmit the preference to its own offspring.

Many other types of nongenetic inheritance that are capable of self-regeneration may involve pathways that are equally complex and indirect. For example, patterns of gene expression can be transmitted nongenetically via "self-sustaining loops."[43] An environmental factor can switch on a particular gene, and the protein produced as a result of that gene's expression can, in addition to its other functions, also maintain that gene in an active (expressed) state. If that same protein is transmitted via the cytoplasm to daughter cells, it can maintain the corresponding gene in an active state in those cells as well, and so on until something intervenes to break the cycle. One particularly widespread and interesting example of self-regeneration is paramutation, whereby a genetic mutation induces a phenotype that perpetuates itself over one or more generations independently of the transmission of the mutant allele.[44] In the case of stably transmitted structural features (like the cellular traits of *Paramecium* and other single-celled eukaryotes, discussed in chapter 5), self-regeneration can involve a template-like process, whereby the synthesis of new structures in the daughter cells is guided by existing structures in the mother cell. The details of these processes remain poorly known. From an evolutionary perspective, the mechanisms of self-regeneration are of interest because their nature determines important properties of nongenetic inheritance, such as the stability of nongenetic transmission across generations and the potential for environmental induction.

EXTENDED HEREDITY

As this outline suggests, we see heredity as comprising more than one channel for the transmission of information across generations—that is, more than one mechanism of inheritance. DNA may be the primary carrier of information in the organism and a primary medium for the transmission of information across generations, but it is certainly not the only medium. In addition to genetic alleles that specify the amino-acid sequences of peptide chains and encode regulatory instructions, organisms transmit nongenetic factors specifying cell structure and polarity, aspects of the epigenome, behavior, and other important traits. Yet, despite the diversity of hereditary mechanisms, we believe that heredity can be conveniently carved up into two slices: *genetic inheritance*, defined as the transmission of DNA sequences (that is, genetic alleles) in the nuclear genome from parents to offspring at conception, and *nongenetic inheritance*, defined as the transmission of other factors (epigenetic, cytoplasmic, structural, somatic, symbiotic, environmental, behavioral) from parents to offspring at conception or during subsequent development. Extended heredity encompasses the totality of these genetic and nongenetic mechanisms of inheritance.[45] This dichotomous scheme is a generalization of "dual inheritance" theory,[46] which recognizes the parallel roles of genetic and cultural inheritance in human populations, to encompass other mechanisms of nongenetic inheritance and extend the scope of extended heredity to nonhuman organisms.[47]

This extended concept of heredity represents a break with past thinking. While twentieth-century biologists regarded the transmission of genes as the sole, universal mechanism of heredity, extended heredity implies the existence of multiple mechanisms of inheritance that operate in parallel, vary in important properties such as stability and potential for environmental induction, and are differentially represented in different taxa, such as plants, animals, and single-celled eukaryotes (figure 2.2). Extended heredity also does not require restricting the definition of "inheritance" to transmission over at least two generations, as some authors have done. While for some purposes the potential for long-term transmission is key, we will see that there are many contexts in which transmission from parents to their offspring alone is very important, and it therefore makes sense to treat all instances of parental influence on the features of descendants under the rubric of heredity.

Genetic and nongenetic mechanisms of inheritance can be thought of as distinct channels for the transmission of biological information across generations.[48] From this perspective, extended heredity is important not because it adds another hereditary channel or set of channels (after all, if we wish, we can regard every chromosome, every locus, or even every nucleotide base-pair position in the genome as a distinct information channel), but because these different channels can convey distinct *types* of information. For example, while genetic inheritance involves random transmission of factors that usually cannot be modified in consistent ways by the ambient environment, nongenetic inheritance involves the transmission of factors that can often respond in consistent ways to environmental conditions. Moreover, while genetic information in the cell nucleus is embodied in linear nucleotide sequences that can be likened to a digital information medium, the factors that we include within nongenetic inheritance range from nucleotide sequences similar to those of nuclear DNA (such as the single-stranded molecules of

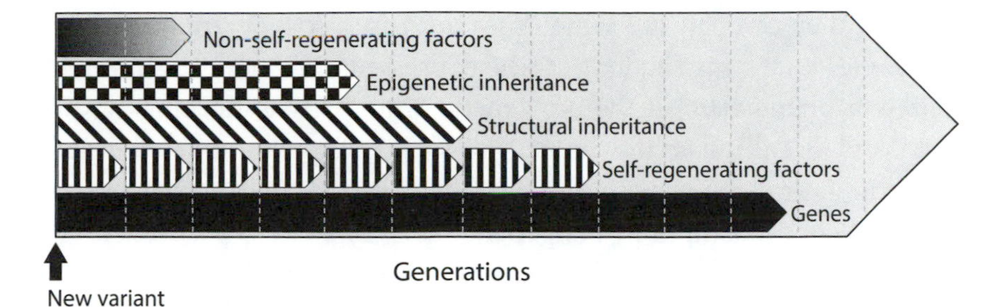

Figure 2.2. Genetic and nongenetic mechanisms of inheritance operate in parallel and vary in properties such as stability and potential to transmit environmentally induced variation across generations. Extended heredity comprises this diverse set of inheritance mechanisms.

RNA) to quantitative, structural, environmental and behavioral forms of variation that embody analogue information.

Richard Dawkins has argued that genes belong to a special class of entities that he called "replicators," defined as things of which copies are made, and whose properties can influence the rate of copying. The only other type of replicator that Dawkins recognized was the meme—an idea or unit of information that can be transmitted between human brains. The existence of nongenetic inheritance means that genes and memes are not the only types of replicators—we could also speak of epigenetic replicators, structural replicators, behavioral replicators, and so on—and extended heredity can be seen as the combined outcome of all these different types of replicators. Theoretical biologists John Maynard Smith and Eörs Szathmáry further distinguished "limited" replicators whose potential range of variation is highly circumscribed (for example, factors that function as developmental or behavioral "on/off switches") from "unlimited" replicators that can take on an essentially infinite variety of states[49]—a useful distinction that we will revisit later on.

Nongenetic inheritance adds a dimension of "phenotypic memory" to heredity—a set of channels whereby accrued environmental influences can be transmitted across generations. Eva Jablonka and Marion Lamb have argued that a genome can be likened to a musical score—a set of instructions that leaves considerable scope for interpretation. According to this analogy, just as two musicians can interpret the same score in different ways that reflect their distinct training and experience (that is, they can play the same notes but produce different music), so the same genetic instructions can be expressed differently (resulting in different phenotypes) if the nongenetic "interpretive machinery" differs. And just as musicians can pass on their interpretation of a musical score to their students without rewriting the notes on the page, so organisms can transmit some nongenetic variation to their offspring independently of variation in DNA sequence.

Extended heredity also suggests a need to revisit the long-held convention of subdividing variation in phenotypic traits among individuals within a population into genetic versus environmental categories. Genetic variation occurs because individuals carry different genes, whereas environmental variation occurs because individuals experience different environments that differentially influence the expression of plastic traits.[50] Under this conventional scheme, offspring-parent resemblance is assumed to result from the transmission of genetic alleles, while the environmental component of phenotypic variation is assumed to be reset between generations and therefore to contribute nothing to heredity. However, in light of nongenetic inheritance, the genetic/environmental dichotomy no longer seems as salient. Indeed, Étienne Danchin, Richard Wagner, and others have argued that it is more useful to categorize phenotypic variation as *heritable* versus *nonheritable*.[51] Under this scheme, heritable variation encompasses both additive genetic variation and those components of nongenetic variation that can be transmitted to offspring. We believe that genetic variation plays a particularly important role in evolution and therefore deserves to be accorded a special status among sources of phenotypic variation. Yet, as we will see, recognizing that nongenetic factors also contribute to heritable variation can allow us to see many evolutionary questions in a new light.

As Tobias Uller and others have argued, nongenetic inheritance also alters the relation between development and heredity.[52] Twentieth-century geneticists and evolutionary biologists were generally happy to treat development as a black box because it was believed that development was under full genetic control, with environmental and stochastic effects merely representing random noise around the genetic signal. Thus, it was felt that the gory details of the developmental processes that link genotype and phenotype could be safely ignored. Extended heredity refutes this idea. Nongenetic inheritance mechanisms influence offspring phenotype by altering gene expression during development, and understanding these mechanisms therefore requires understanding how developmental processes are regulated by both genetic and environmental factors. For example, if we want to understand how children are affected by maternal obesity, there is no escaping the need to understand how embryonic development responds to diet-dependent epigenetic factors transmitted in the egg and to maternal blood glucose levels in utero, or how postnatal development responds to milk composition and maternal behavior. Extended heredity thus brings development back to center stage.

THE IMPLICATIONS OF EXTENDED HEREDITY FOR EVOLUTION

If some nongenetic variation is heritable, then it follows that such variation can respond to natural selection and produce phenotypic change across generations in the absence of genetic change. Such changes do not fit the standard genetic definition of evolution, which is restricted to change in allele frequencies across generations. This definition, devised by evolutionary geneticist Theodosius Dobzhansky, reflected the assumption that genes are the only source of heritable variation, and therefore the only raw material on which natural selection could act to produce phenotypic change across generations. However, it's useful to recall that Charles Darwin was blissfully unaware of the distinction between genetic and nongenetic variation. Darwin's profound insight was that natural selection, applied to heritable variation within a population, can produce change across generations in the average features of organisms because those heritable traits that are consistently associated with a larger number of surviving offspring will be represented in a greater proportion of individuals in each generation.[53] The incorporation of nongenetic mechanisms into heredity does not require any change in this basic Darwinian equation. We will see this more formally in chapter 6, where we incorporate nongenetic inheritance into the equations for evolution.

Given the potential for populations to change over generations as a result of natural selection on nongenetic factors as well as on genes, there is no longer a good reason to restrict the term "evolution" to allele frequency change, and the definition of evolution can be broadened to encompass changes in all heritable traits, whether genetic or nongenetic. Although this may seem like a radical suggestion, it does not actually require a major departure from established terminology; just as the term "cultural evolution" is already widely accepted, so terms such as "epigenetic evolution" could be added to the scientific lexicon to specify the type of hereditary factor involved. Such a broadened definition of evolution would be closer to Darwin's concept of "descent with modification," which was not based on any restrictive assumptions about the nature of heredity.

Yet, as we will see in the next chapter, extended heredity clearly challenges key assumptions of the "neo-Darwinian" Modern Synthesis that developed in the first half of the twentieth century, and that forms the foundation of evolutionary biology to this day. Extended heredity changes the rules governing what can and cannot be transmitted from parents to offspring and how stable the inherited phenotypes are likely to be across generations. Consequently, extended heredity can:

- broaden the range of phenotypic variation on which selection acts
- allow for the maintenance of heritable variation despite persistent directional selection
- avoid Mendelian assortment and recombination
- enable parents to adaptively adjust the phenotype of their offspring
- enable the transmission of acquired pathologies and effects of senescence across generations
- influence the course of evolution via interactions between genetic and nongenetic inheritance
- decouple phenotypic change from genetic change, potentially allowing populations to respond rapidly to natural selection even when genetic variation is lacking
- play a role in rapid coevolutionary chases
- promote reproductive isolation and speciation

To appreciate why biologists can't afford to ignore nongenetic inheritance, consider the roles of heredity at different timescales—long-term "macroevolutionary" phenomena like the buildup and maintenance of lineage-specific features (such as those that distinguish arthropods from vertebrates or humans from chimpanzees) versus the short-term transmission of variation from parents to their offspring. When thinking about microevolutionary questions, such as the benefits of mate choice or the next step in a coevolutionary arms race, the potential importance of nongenetic inheritance is clear because, as we will see, nongenetic inheritance contributes a great deal to a parent's influence on the features of its offspring. When thinking about macroevolution, it's at least plausible to suppose that the conventional, genocentric view could provide a reasonable approximation: by comparing their DNA sequences, biologists from another planet could probably get a good idea of the difference between arthropods and vertebrates and even a first-order idea of the difference between human and chimp. But even in the macroevolutionary context, nongenetic inheritance could still have an important influence on the trajectory of change. For example, the interaction of genes and culture clearly played a major role in human evolution. *Homo sapiens* is a cultural being whose nature cannot be understood in purely physiological terms, so the difference between humans and chimps cannot be fully grasped without taking into account at least this particular nongenetic component of heredity. As we will see, interactions between genes and nongenetic factors could play important roles in the evolution of other organisms as well.

BACK TO THE FUTURE

In this chapter, we attempted to reconstruct heredity from first principles and, in some ways, this exercise brought us back to ideas that were abandoned long ago by mainstream biology. But hasn't the nature of heredity been debated ad nauseam and settled once and for all? Bear with us. As we will see in the next chapter, the development of heredity theory over the past 150 years or so has taken some odd twists and turns that, although perhaps necessary and constructive in their time, ultimately led to a conceptual cul-de-sac. To extricate ourselves from this impasse and move forward, we now need to take a few steps back and ask how we got to this strange place.

3

The Triumph of the Gene

> The proteins of the body cannot induce any changes in the DNA. An inheritance of acquired characters is thus a chemical impossibility.
>
> —Ernst Mayr, *Growth of Biological Thought*, 1982

In this chapter, we trace the path of scientific research and debate that led to the near-universal acceptance of an exclusively genetic model of heredity, and the accompanying denial of the reality of nongenetic hereditary phenomena. We do not doubt that the discovery of the Mendelian gene, the development of genetics, the discovery of DNA's structure, and the synthesis of genetics with Darwinian natural selection in the evolutionary Modern Synthesis were intellectual triumphs of the highest order. Each of these developments is a quintessential example of scientific progress, and each has been tremendously fruitful. Yet, we also believe that Mendelian genes are not the whole story, and that the exclusively genetic concept of heredity was never firmly grounded in evidence or logic. How did scientists arrive at such a model of heredity?

In our view, the train of logic that led to the exclusively genetic concept of heredity can be summarized as follows: First, it was widely assumed that all hereditary phenomena were mediated by a single universal mechanism, and the quest to decipher the laws of heredity became a search for this holy grail. Second, whereas heredity was originally understood as the tendency for some traits to run in families, in the early twentieth century heredity was redefined more narrowly—as the presence of identical "genes" in ancestors and descendants.[54] Third, as knowledge of the transmission and material basis of genes—the Mendelian laws, the structure of DNA, and the genetic code—became more complete, geneticists came to regard nongenetic inheritance as a "chemical impossibility"[55] because they could not imagine how environmental influences could pass from the soma to the germ line, or induce consistent changes in germ-line DNA sequences. This remarkable twist of logic—the rejection of nongenetic inheritance because it was deemed incompatible with the nature of genes—must surely rank among the most influential circular arguments in the history of science. Yet, to be fair, doubts about the plausibility of nongenetic inheritance also stemmed from a lack of compelling evidence in its favor during the early twentieth century, particularly by comparison with the spectacular empirical successes of Mendelian genetics. As always, ideology and politics also appear to have played a role. Strangely, all of these dubious arguments continue to this day to be proffered in defense of an exclusively genetic concept of heredity.

We do not aim to provide a comprehensive treatment of the history of heredity.[56] Rather, we restrict our focus to the factors and developments that led to the triumph of the exclusively genetic concept of heredity during the early years of the twentieth century. Our treatment of this history diverges in important ways from most previous accounts in that, whereas previous works have typically (albeit with notable exceptions[57]) portrayed the history of heredity as a heroic struggle to decipher the Mendelian laws and the structure of the gene, we ask instead how biologists ended up with an overly narrow concept of heredity.

THE DISCOVERY OF HEREDITY

Attempting to imagine how people of long ago used to see the world can be a discomfiting experience. Arthur Koestler likened the worldview of the ancient Babylonians to the magical dreamscape of early childhood,[58] but we need only go back a century or two to encounter a world of disturbingly counterintuitive ideas. Today, we take heredity for granted, and it takes some mental contortions to imagine a biology not grounded in such a concept, but in the late eighteenth and early nineteenth centuries the idea of heredity was only beginning to crystallize, and its nature soon became the subject of heated debate. In 1788 and again in 1790, the Société Royale de Médécine in Paris even offered 600 livres in cash to anyone who could shed light on this enigma.[59]

Before the twentieth century, it was widely assumed that hereditary traits could be acquired during an individual's lifetime through exposure to particular environmental factors or experiences. Such beliefs, which encompassed some real phenomena that we now recognize as nongenetic inheritance but also a variety of superstitions and misconceptions, have been called "Lamarckian" or "soft" inheritance[60] (we will use the latter term to distinguish these ideas from the modern concept of nongenetic inheritance). Belief in soft inheritance has a venerable pedigree.[61] Perhaps the most famous example in Western culture is the biblical story of Jacob and the flock of sheep and goats that he tended for Laban. By agreement, Jacob could add any goats or sheep born with speckled, spotted, or brown coats to his own flock, and therefore invented a devious trick. Jacob tore strips of bark from sticks and placed these sticks at watering troughs where the animals mated. The sight of striped sticks during copulation caused females to give birth to offspring with patterned coats. This ancient story embodies an enduring feature of the popular belief in the inheritance of acquired traits—namely, the assumption of universality. The writer of this part of the book of Genesis seems to have believed that any kind of experience—even a visual one—could have hereditary consequences. Similar notions, such as the idea that a woman's thoughts at the moment of conception could influence the features of her child, persisted in Europe well into the nineteenth century.

In the early nineteenth century, Parisian biologist Jean-Baptiste Lamarck incorporated a form of soft inheritance as a key mechanism in his theory of evolution, which posited that organisms progress through stages of increasing complexity and perfection toward the human form. In his *Philosophie Zoologique* (1809) and other writings, Lamarck argued that the environment influences how organisms behave or use their organs, and that the resulting changes are transmitted to subsequent generations. These ideas are encapsulated in his two "laws of nature":

First Law

> In every animal . . . a more frequent and continuous use of any organ gradually strengthens, develops and enlarges that organ . . .

> while the permanent disuse of any organ imperceptibly weakens and deteriorates it.
>
> **Second Law**
>
> All the acquisitions or losses wrought by nature on individuals, through the influence of the environment . . . and hence through the influence of the predominant use or permanent disuse of any organ; all these are preserved by reproduction to the new individuals which arise.[62]

In modern language, Lamarck's first law embodies the principle of phenotypic plasticity, and his recognition of its importance represents a prescient insight. His second law—the law of the inheritance of acquired traits—became the source of much controversy. Although similar views of inheritance were widespread at the time and Lamarck invoked only the inherited effects of use and disuse, his book ultimately came to be regarded as the flagship manifesto of soft inheritance. Lamarck's belief that organisms specifically acquire and pass on advantageous (that is, adaptive) traits (often called "directed variation") also had a lasting influence; an echo of this belief can be recognized today in the controversial idea that nongenetic inheritance can drive adaptive evolution without the aid of natural selection (a topic that we will revisit in chapter 8).

A number of recent authors have cast the modern concept of nongenetic inheritance as a revival of Lamarckism, but we believe that this is counterproductive. Lamarck was an important thinker, but his two-hundred-year-old writings bear only a vague resemblance to current ideas about heredity and evolution. In fact, in modern usage, the term "Lamarckian" can have at least four different meanings: (1) the theory of evolution that Lamarck proposed; (2) the generation of "directed variation" in response to environment, allowing for adaptation without natural selection; (3) the "genetic encoding" concept of soft inheritance (discussed below); and (4) any instance of the transmission of environmental effects across one or more generations. Meaning (1) is relevant only in a historical sense, meanings (2) and (3) represent highly controversial ideas, while meaning (4) refers to one of the important and widely recognized properties of nongenetic inheritance. The term

"Lamarckian" is therefore prone to sow confusion, and, except in a historical context, we avoid it in this book.

Lamarck did not suggest an explicit physiological mechanism for soft inheritance, but that challenge was taken up fifty years later by a biologist who was born on the other side of the English Channel just as the pages of *Philosophie Zoologique* were being peeled off the printing presses. Charles Darwin rejected Lamarck's teleological and progressive concept of evolution in favor of his own theory based on natural selection of random variation, but he recognized the need for a mechanism of heredity that could account for the resemblance between offspring and their parents. Believing, like almost everyone else at the time, that acquired traits could be passed on to offspring, Darwin proposed a physiological process—the hypothesis of pangenesis—to explain how this could occur:

> According to this hypothesis, every unit or cell of the body throws off gemmules or undeveloped atoms, which are transmitted to the offspring of both sexes, and are multiplied by self-division. They may remain undeveloped during the early years of life or during successive generations; and their development into units or cells, like those from which they were derived, depends on their affinity for, and union with other units or cells previously developed in the due order of growth.[63]

Darwin thus imagined that each cell within the body produces hereditary particles that can self-replicate, travel to the gonads, and enter the gametes. A change in an individual's body that occurred in response to an environmental factor or experience could thereby be transmitted to the offspring.

Notably, like the writer of Genesis, both Lamarck and Darwin assumed that practically any change that occurs in an individual in response to the environment can be passed on to its offspring (albeit with certain caveats, such as Lamarck's belief that such effects were most likely to occur through the mother, or Darwin's belief that mutilations would lead to hereditary changes only if repeated over a large number of generations). As we will see, the problematic assumption that soft inheritance, if it occurs at all, must be universal became a logical foundation for famous experiments later carried out with the aim of refuting soft inheritance.

LIKE BEGETS LIKE

Historian Peter Bowler has argued that belief in soft inheritance followed naturally from the prevailing view of the nature of reproduction.[64] In the nineteenth century, all reproduction was viewed as a kind of budding process, whereby the parental body produces a miniature version of itself from its bodily tissues. Parents literally manufactured their offspring from parts of themselves, and the inheritance of acquired traits followed logically from this concept of reproduction because changes acquired by the parental body (soma) during its lifetime would be transferred automatically to the offspring body formed from parental somatic tissues (figure 3.1). Darwin built this idea into his hypothesis of pangenesis:

> If we suppose a homogeneous gelatinous protozoon to vary and assume a reddish colour, a minute separated particle would naturally, as it grew to full size, retain the same colour; and we should have the simplest form of inheritance. Precisely the same view may be extended to the infinitely numerous and diversified units of which the whole body of one of the higher animals is composed; the separated particles being our gemmules.[65]

In some organisms, reproduction is still viewed in much the same way. Single-celled organisms like bacteria and *Paramecium* divide their cytoplasm to form two daughter cells at each reproductive event. Some simple animals (such as the tiny, tentacled polyps called hydras, familiar from undergraduate biology classes) can reproduce by forming buds that grow into miniature copies of their mother. Many plants can reproduce by forming somatic extensions (ramets) that give rise to complete new plants that are copies of the parent plant. However, in most complex organisms, reproduction ultimately came to be seen as an entirely different kind of process. Rather than being formed from parts of its parents, an offspring came to be seen as an independent entity that developed autonomously by following its own unique hereditary "blueprint," and parents' influence on the features of their offspring was assumed to be limited to the contribution of hereditary units to this blueprint at the moment of conception. This view of reproduction provided an essential foundation for the developing "hard" concept of heredity.

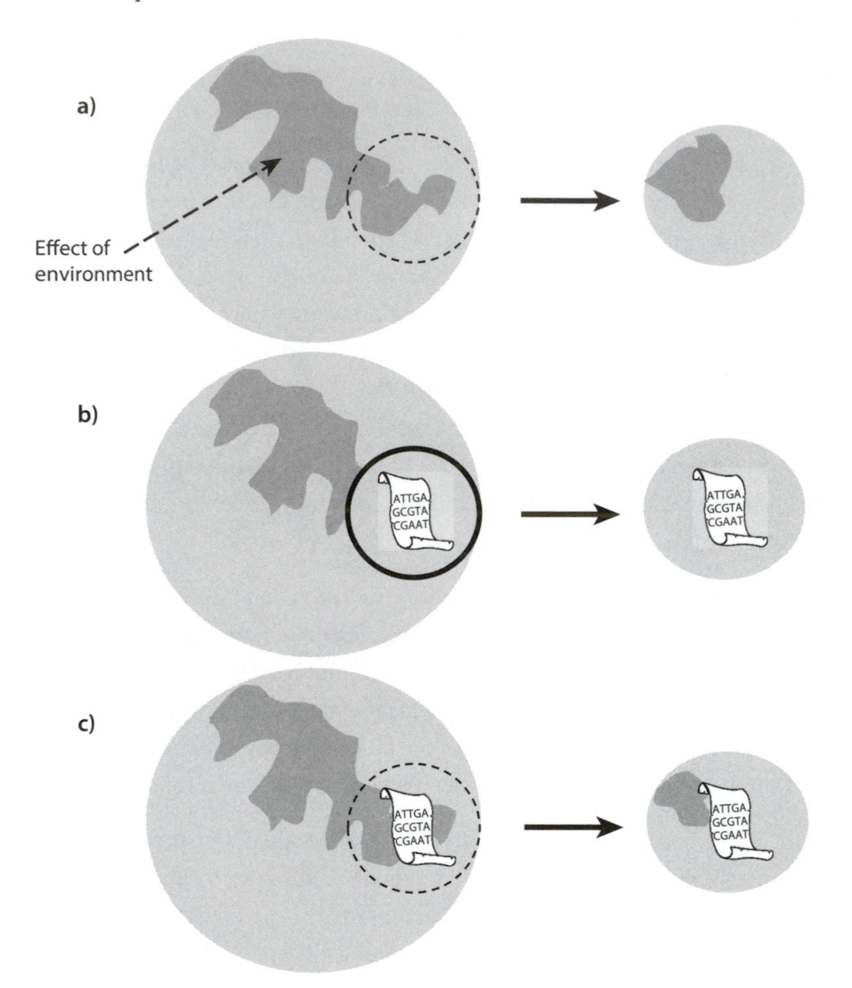

Figure 3.1. Changing views of reproduction and heredity: (a) Prior to the twentieth century, offspring were thought to bud from the maternal body, allowing effects of environment on the mother (dark gray) to influence the offspring. (b) With the advent of genetics, offspring development came to be seen as being controlled by an autonomous genetic blueprint (represented by the scroll), with the "Weismann barrier" (black circle) between soma and germ line precluding transmission of environmental effects from the parental soma to the offspring. (c) Extended heredity incorporates elements of both views, seeing the Weismann barrier (in organisms that possess such a barrier at all) as a porous filter (dashed circle) that allows for some nongenetic influence alongside the effects of the unique offspring genome.

THE HARDENING OF HEREDITY

Darwin believed that hereditary particles could persist within a lineage for many generations even after the organ that originally gave rise to them had disappeared. For example, he used this principle to explain the continued development of a foreskin in Jewish infant boys despite many generations of circumcision.[66] Soon after, some biologists went a step further and posited that the hereditary particles were completely autonomous of the soma and potentially immortal—that is, heredity was "hard." They imagined that these self-replicating particles could pass across many generations unaltered by environmental influences. By assuming a one-way flow of cause and effect from hereditary particle to soma, this hypothesis precluded soft inheritance.[67]

This idea originated with Darwin's cousin, Francis Galton. Beginning in the 1860s, Galton began to argue that parents could only transmit to their offspring such factors as they themselves had inherited from their parents. Galton initially put forth his hypothesis in a series of magazine articles aimed at a broad readership, and based his views on evidence of the inheritance of "genius" in English families. Galton focused on academic and creative achievement rather than mere wealth or social standing, but he neglected the possibility that parents might impart their abilities to their children through teaching, example, and an environment conducive to learning, rather than simply via a hereditary substance transferred at conception. Indeed, Galton believed that human ability is largely determined by "nature" (that is, hereditary factors), which he regarded as far more important than "nurture" (that is, the developmental environment).[68] In an article written in 1865, Galton used a striking metaphor to express the autonomy and continuity of hereditary factors, foreshadowing the modern concept of heredity as the transmission of potentially immortal genes:

> We shall therefore take an approximately correct view of the origin of our life, if we consider our own embryos to have sprung immediately from those embryos whence our parents were developed, and these from the embryos of their parents, and so on for ever.[69]

Galton's writings also sowed the seeds of an enduring misconception—the idea of "nature" and "nurture" as a mutually exclusive dichotomy. This idea persists in the popular view that traits that have a genetic basis must be impervious to environmental influence (that is, plasticity), giving rise, for example, to the belief that intelligence is fixed at conception and unalterable by upbringing and education. Yet, a basic and noncontroversial premise of genetics is that the heritability of a trait (that is, the extent to which the trait "runs in families") bears little if any relation to its plasticity (that is, the extent to which the trait's expression within a given individual can be influenced by the environment); if the environment changes, even a strongly heritable trait can be expressed very differently.[70] This principle is nicely illustrated by the striking increase in average human height over the past century.[71] Height is among the most heritable of human traits, meaning that children tend to be very similar to their parents in height when both grow up in the same environment. However, height also responds very strongly to changes in environmental factors such as diet. This allows average height to track changes in environment across generations and explains why child migrants to richer countries often grow much taller than their parents.

Galton's writings on hard heredity influenced the German biologist August Weismann, whose 1893 synthesis of the rapidly expanding knowledge of cell division, gamete formation, and the process of fertilization established a physiological foundation for an exclusively hard model of heredity.[72] Weismann recognized the chromosomes as bearers of hereditary factors, which pair up and then separate as complete sets into the gamete cells in the process of meiosis. At fertilization, the chromosome sets from egg and sperm unite in the zygote, thereby conveying complementary sets of maternal and paternal hereditary factors to the offspring. Weismann also reasoned that the germ line (or "germ plasm") represents a kind of bottleneck between generations, allowing for the passage of hereditary factors only through a single, specialized cell. Because, in some complex animals such as vertebrates and arthropods, the germ line is "sequestered" early on in development, and because almost none of the parental body is transmitted to the offspring along with the hereditary factors, Weismann argued that changes in bodily (somatic) tissues in response to environmental factors, use and disuse of organs, or mutilations cannot affect the hereditary factors contained in the germ line, and therefore that acquired traits cannot be transmitted to descendants. The apparent boundary between the somatic tissues and the germ line, which came to be known as the "Weismann barrier," is still widely regarded as a key impediment to nongenetic inheritance. However, such a barrier is lacking in many organisms; for example, there is no distinct germ line in single-celled organisms, while plants produce new germ line from somatic tissue on each branch and many animals (such as mollusks, annelid worms, and echinoderms like sea urchins) generate germ line tissue from adult somatic tissues or multipotent stem cells.[73] Indeed, as we will see, the barrier between soma and germ line appears to be quite porous even in animals that sequester the germ line during embryonic development. Weismann's theory paved the way for the development of genetics and furnished a key rationale for the rejection of soft inheritance.[74]

Prior to the twentieth century, heredity was understood as the like-begets-like property of reproduction—the process that conferred a resemblance between offspring and their parents and caused traits such as illnesses, facial features, or temperament to "run in families." Heredity was thus defined in phenomenological rather than mechanistic terms—as the substance and pattern of familial resemblance rather than a chemical process. But the early twentieth century saw an important semantic shift. In a landmark paper published in 1911 in the *American Naturalist*, Danish botanist Wilhelm Johannsen redefined heredity as "the presence of identical genes in ancestors and descendants." He also distinguished the "genotype" (the sum of the genes possessed by an individual) from the "phenotype" (the mortal body shaped and controlled by the expression of genes).[75] The quest for the universal mechanism of heredity was thus transformed into a search for the nature of chemical factors called "genes." Johannsen's concept of the phenotype as the bodily features that are shaped by heredity but whose properties cannot alter the hereditary units was a direct extrapolation of Weismann's theory of heredity as the continuity of the "germ plasm," which was itself an abstraction of Galton's chain of embryos. By Johannsen's definition, only genes were transmitted across generations, and, although the physical nature of the gene was unknown, genes were assumed to be impermeable to environmental influence.

ABSENCE OF EVIDENCE OR EVIDENCE OF ABSENCE?

Starting in the late nineteenth century, the science of heredity was revolutionized by the rapid development of an empirical research program employing laboratory cultures of small, rapidly breeding animals, plants, and other organisms as experimental models in the search for the universal mechanism of heredity. This empirical revolution was spurred, in part, by the desire to test the competing hard and soft models of heredity.

The most important early champions of the hard heredity concept, Francis Galton and August Weismann, also made pioneering contributions to the empirical study of heredity. In the 1870s, Galton performed a series of experiments with rabbits to test Darwin's pangenesis theory of soft inheritance. Reasoning that gemmules must be present in the blood, Galton carried out blood transfusions from several distinct breeds of rabbits into a reference breed called "silver-gray," which is characterized by an even gray coat and distinctive ear shape. He found that, despite large transfusions of blood, his silver-gray rabbits reliably produced offspring with characteristic silver-gray features. From his null results for coat color and ear shape, Galton inferred that the blood transfusions had no transgenerational effects of any kind, and declared in a paper published in the *Proceedings of the Royal Society of London* that he had refuted pangenesis.[76]

Weismann set out in 1887 to disprove the inheritance of mutilations and thereby show that "there are no direct proofs supporting the Lamarckian principle." This putative form of soft inheritance was supported by dubious anecdotal observations of parental injuries that reappeared in the offspring, such as "a bull which had accidentally lost its tail [and then] begat tailless calves." Weismann reasoned that, if mutilations can be inherited in the next generation, or after a few generations of repeated mutilation, then severing the tails of mice over five generations would lead to a reduction in tail length at birth. Not seeing any change in tail length, Weismann concluded that his experiment refuted anecdotal claims of heritable mutilation such as the case of the tailless bull. He also pointed out that cases of repeated mutilation in humans (such as circumcision and foot-binding) or other animals (such as the customary removal of the tail in certain breeds of sheep) showed that even mutilations that recurred over many generations produced no visible effect on offspring features. To Weismann, such evidence deprived Lamarckism of one of its key claims to empirical support.[77]

In hindsight, it's easy to spot the flaws in these experiments. Both Galton and Weismann based their inferences on the venerable but unfounded assumption of universality—the notion that soft inheritance, if it occurs, must be detectable by means of any kind of manipulation of parental features, including blood transfusion and mutilation. Neither Galton nor Weismann attempted to investigate the effects of parental diet, social environment, rearing temperature, or any of the other environmental manipulations that have since been shown to influence offspring features in many species. Moreover, the limited experimental and statistical tools available at the time would have made it very difficult in any case to carry out such studies. As we will see, similar technical and inferential problems also dogged many early twentieth-century studies on soft inheritance.

Between 1901 and 1907, Wilhelm Johannsen carried out a study that marked a turning point in the history of genetics, and is still often cited as evidence of the exclusively hard nature of heredity. Johannsen established nineteen "pure lines" from individual bean plants that were allowed to self-fertilize. Because self-fertilization represents an extreme form of inbreeding, it eventually results in genetically homogeneous lines. Johannsen then applied artificial selection for bean weight within each of the nineteen lines and found that there was no clear relation between maternal and offspring bean weight—that is, bean weight was nonheritable and thus did not respond to selection. Based on these results, Johannsen concluded that "selection . . . is effective only in so far as it selects out representatives of an already existent genotype,"[78] whereas variation of a purely environmental nature is nonheritable. These results provided key empirical support for Johannsen's genotype/phenotype dichotomy.[79]

Some experiments also provided evidence of soft inheritance.[80] For example, between 1906 and 1915, Francis Sumner reared mice in either warm or cold temperatures and assessed the impact of rearing conditions on the expression of a series of morphological features in the exposed mice and their offspring. He found that cold-exposed mice grew shorter tails, ears, and feet, and also produced offspring that expressed these traits even when reared in warm conditions—a finding that would be interpreted today as an adaptive parental effect. Some biologists also viewed the effects of maternal alcohol consumption on offspring as evidence of soft inheritance (a fascinating story in its own right that we will revisit in chapter 10).

Embryologists also took a keen interest in the problem of heredity and, for a while, played a leading role in efforts to identify its physical location within the cell. The key question was whether heredity resided exclusively in the nucleus or also in the cytoplasm. The German embryologist Theodor Boveri came up with an ingenious way to get at this tricky problem by taking advantage of a peculiar feature of sea urchin biology—the ability of haploid zygotes to develop into embryos. Boveri de-nucleated the eggs of sea urchins and fertilized them with sperm from a different species with a distinct pattern of embryonic development. Although the results of Boveri's early experiments were inconclusive and controversial,[81] they suggested that early phases of embryonic development resembled those of the egg-donor species and were therefore controlled by the cytoplasm. Later embryological investigations revealed that the egg cytoplasm is highly structured, containing molecular gradients and complex subcellular architecture. When the egg begins to divide, the developmental fate of the early embryonic cell lines is determined not by the genes in the nucleus but by the cytoplasmic components that the cells receive. A key element of the egg's cytoplasmic architecture is its polarity, with the axis between its "animal" and "vegetal" poles becoming the long axis of the developing embryo. Modern research has identified many other aspects of early embryonic development that are controlled by the egg cytoplasm's content and structure, and it's now clear that the cytoplasm plays a key role in development at least until the point when the embryo's own genome begins to be expressed.[82] This work also led to the discovery that mitochondria and chloroplasts (cytoplasmic organelles descended from bacterial symbionts) possess their own genomes, and that these cytoplasmic genes can affect important metabolic traits.[83] However, geneticists countered that other features of the cytoplasm were determined by nuclear genes expressed in the mother's body, and therefore that such cytoplasmic factors could not be regarded as autonomous bearers of hereditary information. As we will see in chapter 5, the potential for cytoplasmic factors (other than mitochondria and chloroplasts) to play an independent hereditary role remains an open question to this day.

However, none of the early evidence was sufficiently compelling to challenge the growing conviction that heredity was hard, and researchers who continued to search for evidence of soft inheritance were increasingly regarded as misguided obscurantists, or worse. Notably, the growing doubt cast on soft inheritance was reinforced by a notorious case of alleged scientific fraud. In 1909, Austrian zoologist Paul Kammerer published a study purporting to show that the largely terrestrial midwife toad, when reared in an aquatic environment for several generations, acquired morphological and behavioral features associated with aquatic breeding—results that Kammerer interpreted as evidence of soft inheritance. Kammerer's work initially attracted a lot of interest, but some of his own laboratory assistants (among them his paramour, Alma Mahler, illustrious Viennese socialite and recent widow of the famous composer-conductor Gustav Mahler) claimed that his experimentation was sloppy and even fraudulent. Ultimately, Kammerer was accused of falsifying his results and committed suicide soon thereafter. Whether or not Kammerer actually committed fraud,[84] his tragic story did not help the concept of soft heredity.

Over the next two decades, research on soft inheritance shifted increasingly to single-celled organisms. German zoologist Max Hartmann and his protégé, Victor Jollos, showed that environmentally induced changes in phenotypic traits such as heat resistance could be transmitted over many cell generations in *Paramecium*. Jollos called such changes "Dauermodifikationen" and believed that they represented heritable adaptive changes in response to environmental stresses. But work was interrupted by the rise of the Nazis. Jollos (who was of Jewish descent) sought refuge in the United States but was shunned by American geneticists because of his "Lamarckian" ideas and could not secure an academic position.[85]

FINDING THE HOLY GRAIL

While research on soft inheritance produced complex and inconclusive results, the rediscovery of Gregor Mendel's famous study on peas inspired a highly successful research program. Mendel was a Moravian monk who carried out breeding experiments with pea plants in the garden of his monastery, crossing strains that varied in traits such as color and texture and counting the numbers of offspring of each type that resulted from his crosses. Based on his observations, he suggested that each individual carries two determinative factors for each trait (for example, one individual might carry green/green factors for color, while another might carry yellow/green), but transmits only one of these factors to each of its progeny. Some factors are dominant to others; for example, an individual that carries yellow/green will be yellow because the yellow factor is dominant to the green factor. However, a factor retains its properties across generations, so an offspring of two yellow parents that each carry a green factor will be green if it inherits a green factor from each parent.

Mendel published his findings in 1865, but they were ignored for thirty-five years. Yet, within a few years of the rediscovery of Mendel's paper in 1900, researchers demonstrated the operation of Mendel's "laws" in plants, insects, birds, mammals, and even single-celled microorganisms. Here at last was a widespread pattern of inheritance that could be readily reconciled with the theory of hard heredity. The most important work was done by Thomas Hunt Morgan, whose famous "fly lab" at Columbia University in New York City demonstrated Mendelian inheritance of a large number of mutations that could be detected by the presence of visible traits such as white eyes, kinky wings, or missing bristles in the tiny vinegar fly *Drosophila melanogaster*. Through meticulous analysis of fly pedigrees, Morgan painstakingly verified Mendel's laws of heredity, extended these laws to sex-linked inheritance (that is, the transmission of traits determined by genes residing on the X chromosome), and even mapped the physical locations of genes on chromosomes. Morgan trained a generation of leading geneticists and went on to win the Nobel Prize.

At the same time, Morgan and other champions of the new science of genetics sought to cleanse biology of all vestiges of soft inheritance—an idea that they regarded as a pseudoscientific holdover from the nineteenth century that now served only to impede scientific progress. In Morgan's words, "The theory of the inheritance of acquired traits had obscured for a long time all problems dealing with heredity."[86] And the geneticists triumphed; well before the middle of the twentieth century, leading biologists in the United Kingdom and the United States were nearly unanimous in their belief that the universal mechanism of heredity was embodied in Mendelian genes whose nature was impervious to environmental influence.

Importantly, a key guiding assumption throughout the search for the physical basis of heredity in the late nineteenth and early twentieth centuries was the existence of a single, universal mechanism that could account for all forms and instances of inheritance. This assumption was clearly implicit in the language used. For example, in 1919, Morgan titled his book on Mendelian genetics *The Physical Basis of Heredity*.[87] An obvious corollary of the belief in the existence of a single, universal mechanism of heredity was that there was no point in searching for other mechanisms, and by the 1920s, most prominent biologists were convinced that the holy grail had been found.

DECONSTRUCTING AN EMPIRICAL FAILURE

From our vantage point, it's obvious that geneticists' categorical rejection of nongenetic inheritance was an inferential error—a premature conclusion reached on the basis of flawed evidence. To see this, consider the nature of the evidence sought. At a symposium on the inheritance of acquired traits organized by the American Philosophical Society in 1923, Frank Hanson of Washington University defined the problem thus:

> Can a structural change in the body induced by some change in use or disuse, or by a change in surrounding influence, affect the germ cells in such a specific or representative way that the offspring will through its inheritance exhibit, even in a slight degree, the modification which the parent acquired?[88]

In other words, for early researchers working on the nature of heredity, the fundamental question was whether parental exposure to a particular environment or experience could have consistent effects on offspring traits. The broad consensus that emerged, at least among leading biologists in the United Kingdom and the United States, was that such effects simply did not occur. Yet, as we will see in chapters 4 and 5, effects that fit Hanson's definition have now been observed in numerous studies on a variety of animals, plants, and other organisms, and even Johannsen's famous work on "pure lines" has been superseded by experiments showing that genetically homogeneous (isogenic) lines can harbor heritable variation. Early studies therefore failed to obtain convincing evidence of a widespread class of phenomena whose existence is now supported by a large scientific literature.

Admittedly, some biologists proposed more stringent definitions. For example, microbial geneticist Joshua Lederberg wrote in 1948 that "heritable variation should be defined as a change of type which, after removal of the agent provoking the change, retains the capacity for indefinite propagation."[89] Lederberg's definition would disqualify some forms of nongenetic inheritance, such as effects of parental environment that wane over one or two generations, from the definition of heredity. However, some highly stable nongenetic factors (such as the cortical structure variants studied by biologists working on *Paramecium* and other protists, discussed in chapter 5) would surely pass Lederberg's test. Moreover, many nongenetic factors have the property of self-regeneration, and such variants would fit Lederberg's definition because they can be propagated indefinitely. The question of "capacity for indefinite propagation" therefore does not suffice to cleanly differentiate hard and soft inheritance and was not pivotal to the controversy.

Had unequivocal evidence of parental effects been obtained a century ago, the subsequent history of genetics and evolutionary biology might have unfolded differently. So how did this empirical failure come about?

This historical sketch suggests several reasons why early studies failed to provide compelling examples of soft inheritance, leading to the conclusion that soft inheritance had been refuted. First, as we saw previously, most empirical researchers appear to have assumed that all hereditary phenomena would ultimately be explained by a single, universal mechanism. The empirical study of heredity was thus seen as a contest between two mutually exclusive alternatives. Second, the Mendelian-genetic research program had important advantages over attempts to demonstrate soft inheritance. Mendelian inheritance could be demonstrated by examining the transmission within pedigrees of readily observable mutant phenotypes such as crinkly wings or white eye color—a powerful paradigm that allowed early geneticists to rapidly confirm Mendel's laws in diverse species. By contrast, detecting nongenetic inheritance required complex experiments involving the manipulation of parental environment or phenotype and quantification of variable effects on descendants at a time when scientists still lacked many of the basic tools for such work.[90] It's hardly surprising that leading biologists embraced the Mendelian mechanism and rejected soft inheritance.

To appreciate the difficulties early researchers faced, it's useful to consider how studies on nongenetic inheritance are conducted today, and the nature of the evidence that such studies generate. Modern research has shown that the role of nongenetic inheritance is often seen in continuously varying ("quantitative") traits, and especially in behavior, physiology, and life history. Although such effects can be considerable, their detection often requires sensitive phenotypic assays and sophisticated statistical approaches that only became available in the second half of the twentieth century. And, of course, the study of the molecular basis of nongenetic inheritance only became possible with the advent of molecular biology in the late twentieth century, removing theoretical objections by revealing the existence of cellular mechanisms capable of mediating such effects. Importantly, modern researchers also recognize that not all features are heritable, either genetically or nongenetically. In other words, early twentieth-century biology was simply unprepared for the study of nongenetic inheritance.

REFUTATION BY REDEFINITION

The categorical rejection of soft inheritance also relied on an odd turn of logic—the redefinition of soft inheritance as "genetic encoding." As we saw previously, early geneticists widely assumed that heredity must depend on a single, universal mechanism. As evidence of Mendelian inheritance in both plants and animals began to accumulate in the first decades of the twentieth century, geneticists naturally concluded that they had at long last found the key to the mystery of heredity. But, importantly, the new Mendelian-genetic concept of heredity led to a reformulation of the debate on soft inheritance as well. Seeing the world through the prism of the new ideas, early geneticists concluded that soft inheritance, if it occurred, would also have to be mediated by the transmission of genes, and would therefore have to involve the precise and consistent modification of germ-line genes by the environment.[91] For example, if the stress experienced by a parent were to influence the phenotype of its offspring in a consistent way (so that, let's say, severely

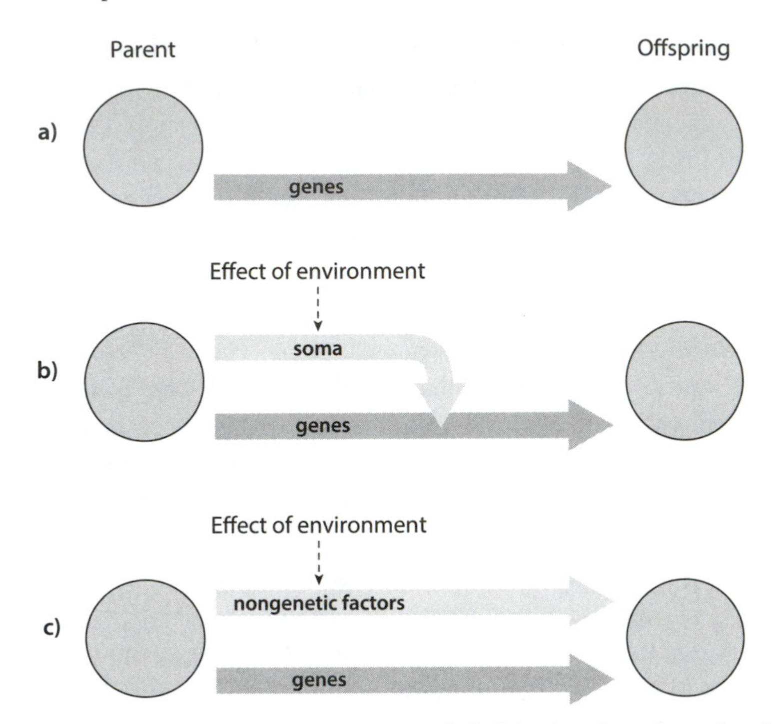

Figure 3.2. (a) Twentieth-century geneticists concluded that heredity is limited to the transmission of genes from parent to offspring. (b) They also believed that, for "soft inheritance" to occur, environmental influences would have to induce specific and repeatable modifications of genes in the germ line. "Soft inheritance" was deemed impossible because geneticists were unaware of any molecular mechanisms that could bring about such "genetic encoding" of environmental influences. (c) By contrast, the modern concept of nongenetic inheritance posits the transmission of nongenetic factors alongside genetic alleles and therefore does not require genetic encoding. (Modified from Bonduriansky, "Rethinking Heredity, Again," 2012.)

stressed parents tended to produce smaller or sicklier offspring), it was assumed that the parent's experience of stress would have to somehow modify the structure of genes in the parental germ line. This concept of soft inheritance was articulated with ever-increasing precision as the physical nature of genes was uncovered, such that, by the second half of the twentieth century, when genes were defined as segments of DNA embodying hereditary information in a genetic code, it was routinely asserted that soft inheritance would have to involve specific changes in the base-pair sequence of germ-line DNA so as to produce specific changes in offspring phenotype (box 3.1; figure 3.2). Because biologists knew of no molecular mechanism that could bring about such "genetic encoding" of environmental influences, soft inheritance was dismissed as a "chemical impossibility."[92]

Box 3.1. Development of the "Genetic Encoding" Idea in the Words of Leading Geneticists[93]

August Weismann (1893) "At the present day I can therefore state my conviction . . . that neither injuries, functional hypertrophy and atrophy, structural variations due to the effect of temperature or nutrition, nor any other influence of environment on the body, can be communicated to the germ-cells, and so become transmissible."

Wilhelm Johannsen (1911) "Personal qualities are then the reactions of the gametes joining to form a zygote; but the nature of the gametes is not determined by the personal qualities of the parents or ancestors in question."

William Bateson (1915) "There is now scarcely any doubt that the germ-cells of which the offspring are composed possess from the beginning ingredients determining their powers and attributes; and that, with rare and doubtful exceptions, it is not in the power of the parent, by use, disuse, or otherwise, to increase or diminish this total."

Thomas Hunt Morgan (1926) "The Lamarckian theory of the inheritance of acquired characters . . . postulates [that] the germ-cells are affected by the body in the sense that a change in a character may bring about corresponding alterations in specific genes."

J.B.S. Haldane and Julian Huxley (1934) "All acquired characters . . . affect the soma. But how is a change in the soma to alter the germplasm?"

Julian Huxley (1949) "The observed facts about reproduction and the chromosome mechanism of inheritance make it extremely difficult to see how a somatic effect (say of sunlight on the colour of our skins) could find its way into the elaborately self-regulating system of self-reproducing genes."

Continued on page 44

Theodosius Dobzhansky (1951) "The error of the Lamarckian belief in the inheritance of acquired characters is due to a failure to recognize that the phenotype . . . is a by-product of the gene reproduction . . . and not vice versa."

Francis Crick (1966) "Notice that as far as we know the cell can translate in one direction only, from nucleic acid to protein, not from protein to nucleic acid. This hypothesis is known as the Central Dogma. . . . detailed work on the genetic code . . . illuminates such concepts as the absence of the inheritance of acquired characteristics."

Theodosius Dobzhansky (1970) "Consider such an acquired trait as big muscles strengthened by exercise. Its inheritance would require that some product secreted by the muscles changed the nucleotide sequence or number in the DNA chains of some genes. Such changes are unknown and seem quite improbable."

Ernst Mayr (1982) "The proteins of the body cannot induce any changes in the DNA. An inheritance of acquired characters is thus a chemical impossibility."

In hindsight, it's clear that the "genetic encoding" concept was based on a peculiar form of circular reasoning. Heredity was redefined as the transmission of Mendelian genes, and every new discovery about the nature of these genes was assumed to further circumscribe the potential for soft inheritance. Recycling an analogy made decades earlier by August Weismann, Julian Huxley derisively compared soft inheritance to the idea that a telegram sent in Chinese could arrive at its destination spontaneously translated into English.[94] Yet, Huxley's dismissal of soft inheritance amounted to the argument that a statement such as "I would like a pizza" cannot be made in Chinese because that language lacks subjunctives and indefinite articles. This logic breaks down as soon as it's recognized that, just as English grammar is not the only possible basis for communication, genes are not the only factors transmitted across generations, and need not be the sole bearers of heredity. Indeed, as we saw in chapter 2, the universal nature of cellular reproduction involves the transmission of other factors alongside genes, providing a number of channels for nongenetic transmission of information across generations.

The rapid and unequivocal acceptance of the Galtonian concept of heredity by Western biologists suggests that ideological commitments may have played a role as well. While ideology and politics may seem irrelevant to the outcome of a scientific debate, the ideological positions staked out by proponents of competing visions of heredity were regarded at the time as thoroughly grounded in scientific theory and evidence, and we believe that they are also a fascinating and important part of the story. We briefly explore this subplot in the next section.

HEREDITY AS IDEOLOGY

Ideology is clearly evident in the writings of the earliest advocate of hard heredity, Francis Galton, and even more so in the views of some of his followers. As we have seen, Galton believed that hereditary and environmental influences—nature and nurture—are mutually incompatible, and an obvious corollary of his views was that little could be done to help the less fortunate members of society, because their lot in life was an inescapable result of their poor hereditary endowment. On the basis of this belief, Galton called for government intervention in human breeding with the goal of improving the human hereditary stock through the elimination of undesirable individuals from the breeding pool and the encouragement of reproduction by those considered to embody the best traits. While Galton believed that breeding by undesirable individuals could be prevented "with little severity" through the encouragement of celibacy, he argued that if such people "continued to procreate children, inferior in moral, intellectual and physical qualities, it is easy to believe the time may come when such persons would be considered as enemies of the State, and to have forfeited all claims to kindness."[95] Galton thus laid the foundations of the eugenics movement.

In the first half of the twentieth century, eugenics was a respectable intellectual position, with prestigious scientific journals devoted to its study (you can still leaf through their yellowed pages in the archival stacks of university libraries), lavish funding from philanthropic agencies such as the Carnegie and Rockefeller Foundations, and the enthusiastic advocacy of leading scientists such as W. Bateson, R. A. Fisher, and H. J. Muller. The eugenics movement also spawned a highly successful political program that was embraced in some form by governments in countries such as the United States, Canada, Sweden, and Japan and, in some cases, implemented in the form of mass sterilization of people of "bad heredity," such as alcoholics, the mentally ill, orphans, "sexual perverts," convicts, prostitutes, or individuals with learning disabilities. The eugenics program reached its apogee in Germany with the mass murder of homosexuals and the physically and mentally disabled and, ultimately, the Nazi campaign of genocide against "undesirable races." Tainted by Nazi atrocities, eugenics fell out of favor after World War II. Eugenics journals were renamed or shut down (for example, the *Annals of Eugenics* was reborn as the *Annals of Human Genetics*, while the *Eugenics Review* simply closed). Today, even the word is all but forgotten.[96]

Galton's intellectual legacy thus encompasses both the scientific doctrine of hard heredity and the ideology of eugenics. These ideas were closely linked, and it's quite possible that the hard heredity doctrine owed some of its early success to the popularity of eugenics. After all, hard heredity allowed social elites to see themselves as intrinsically better rather than simply luckier than their less fortunate compatriots. Hard heredity also provided a putative scientific foundation for prevailing racist and anti-redistributionist attitudes.[97] Eugenics and overt racism eventually fell out of fashion in Western intellectual discourse. Yet, the scientific standing of hard heredity was not seriously tarnished by its unsavory history.

By contrast, the reputation of soft heredity was dealt a severe blow by the rise of Lysenkoism in the USSR during the 1930s and '40s. Communist ideology is in some respects inherently conducive to a belief in soft heredity. If hereditary factors can be modified through engineered changes in the environment within which people live, then society and people can be molded by the Communist Party in accordance with the principles of Marxism-Leninism. Stalin himself expressed a vague belief in "neo-Lamarckism" in his early work *Anarchism or Socialism?*[98] (echoing the nineteenth-century writings of Friedrich Engels), and such ideas may have influenced his ham-fisted attempts at social engineering.

The Soviet political system promoted ruthless careerists who knew how to speak the language of ideological orthodoxy, enabling a scientifically illiterate Ukrainian agronomist named Trofim Lysenko to rise to prominence and power and eventually become a member of the Soviet Academy of Sciences and head of leading research institutes. Lysenko rejected genetics (which he derisively referred to as "Mendelism-Morganism") as a false science contrived to serve the interests of Western capitalism and citied Russian and Western geneticists' support for eugenics as evidence of the corrupt, reactionary nature of genetics. In place of genetics, Lysenko promoted a pseudoscientific notion of soft inheritance that allegedly even allowed for the transformation of one species into another. In a country brought to the brink of starvation by Stalin's forced collectivization of agriculture, Lysenko promised to achieve miraculous improvements in harvests of wheat and other crops by applying vaguely defined techniques based on environmental manipulation of hereditary factors. With Stalin's backing, Lysenko purged genetics research and teaching from Soviet universities and institutes.[99]

Lysenkoism made no positive intellectual contribution to biology, but it incited a backlash in the West against research on soft heredity.[100] The destruction of Soviet genetics shocked Western biologists. Leading geneticist Theodosius Dobzhansky had declined an invitation by his Russian colleague Nikolai Vavilov to return from the United States to the USSR, only to learn a few years later of Vavilov's death in prison. Hermann J. Muller, an American communist who spent years working in Leningrad during the 1930s, ultimately left the USSR in disgust.[101] Julian Huxley, a prominent architect of the evolutionary Modern Synthesis, eloquently vented his outrage in his 1949 book *Soviet Genetics and World Science*,[102] a work that, not coincidentally, contains a particularly scathing critique of soft inheritance. Others, like the politically conservative American botanist Conway Zirkle, used Lysenko as a pretext to launch attacks on leftist academics. In fact, during the postwar McCarthyist hysteria in the United States, perceived sympathy for "Lysenkoist" ideas such as the inheritance of acquired traits could be grounds for dismissal from university positions.[103]

Ironically, the communists' ideological position was in some ways a mirror image of that of the eugenicists. In both ideological camps, hard heredity was generally seen as a doctrine that precluded the possibility of environmental intervention in human traits and, for this reason, it was rejected by the former and embraced by the latter.[104] Yet, both positions rested on a misconception—the Galtonian fallacy that nature precludes nurture. In reality, it's quite clear that the potential for environmental intervention in human development is largely independent of heredity; whatever its genetic or nongenetic hereditary endowment, a child is likely to benefit from a healthy, stimulating, and nurturing environment. This fact, exemplified by the combination of high heritability and strong plasticity of human height, is even more apparent in human cognitive development.

THE MODERN SYNTHESIS AND ITS CREATION MYTH

The rediscovery of Mendel's work led to a monumental breakthrough in the study of heredity that soon culminated in the new field of genetics. Some early twentieth-century biologists initially regarded Mendelian genes as incompatible with Darwin's theory of evolution, seeing these factors as evidence that evolution was not an incremental process driven by natural selection of continuously varying traits but, rather, a series of more abrupt changes guided by the inherent properties of mutation and the appearance of discrete macrovariations.[105] Eventually, however, a number of prominent biologists—J.B.S. Haldane, R. A. Fisher, Sewall Wright, Theodosius Dobzhansky, Julian Huxley, and others—recognized that natural selection on random mutations in Mendelian genes could readily generate incremental phenotypic change in populations, and this reconciliation of genetics with Darwin's theory of evolution by natural selection became known as the Modern Synthesis of evolutionary biology.[106] Geneticists quickly realized that evolution could be analyzed mathematically, using a series of techniques that came to be known as population genetics. This approach, whereby evolution is modeled as changes in the frequencies of alleles that segregate according to Mendelian rules, has been used to explore a great variety of evolutionary questions and can be regarded as the fundamental paradigm of modern evolutionary biology.

In its typical form, population genetic analysis embodies several key assumptions about the nature of heredity: that genes are the only heritable factors, that all heritable variation segregates in accordance with Mendelian rules, that mutation is rare, and that environmental induction of hereditary variation does not occur. To justify these assumptions, geneticists settled on an apocryphal history of their science where the impossibility of nongenetic inheritance was supposedly demonstrated by numerous empirical investigations, and the reason for its nonexistence became clear once it was realized that genes are the sole bearers of heredity. Prominent geneticists Lesley Dunn and Theodosius Dobzhansky put it thus in 1946:

> For at least thirty years, many biologists, among whom the names of Weismann and Galton should be mentioned most prominently, made studies on plants and animals to see whether acquired characters are ever inherited. The number of believers in this contingency dwindled steadily, because the outcome of the experiments on inheritance of acquired characters was so uniformly negative. The rediscovery of Mendel's laws and the development of genetics in the current century have produced a much better understanding of the mechanism of heredity. In the light of this understanding we can clearly see why acquired characters are not inherited. This is because heredity is transmitted through genes.[107]

As we've seen, this historical narrative is at least partly a myth. The discovery of the Mendelian gene did not preclude nongenetic inheritance—it merely led to the marginalization of research on nongenetic inheritance, and to its exclusion from evolutionary theory. But this myth has been retold in innumerable texts and undergraduate courses, and continues to influence biologists' thinking to this day.[108]

As we argued in chapter 2, we believe that the genetic concept of heredity at the heart of the evolutionary Modern Synthesis is incomplete. Although genes play a central role in heredity, in all cellular life-forms, heredity also involves the transmission of nongenetic factors. Every individual's unique genetic blueprint begins its career inside a cell made by its mother's body, and this cell's contents and structure (along with paternal contributions borne in the semen) guide the initial phases of embryonic development. In many species, embryonic development proceeds for an extended period inside the mother's body, and the parent-offspring association sometimes continues well beyond birth. In a sense, the expression of the offspring genome during development is therefore overlaid on a parental phenotypic scaffold. In the next two chapters, we will explore some examples of how nongenetic parental influences can shape offspring development.

4

Monsters, Worms, and Rats

> Discovery commences with the awareness of anomaly, i.e., with the recognition that nature has somehow violated the paradigm-induced expectations that govern normal science.
>
> —Thomas Kuhn, *Structure of Scientific Revolutions*, 1970

We now examine some of the evidence that underpins the ideas in this book. Our focus in this chapter is on epigenetic inheritance, and we will start by attempting to clarify the meaning (or rather, meanings) of this increasingly fashionable term. In the next chapter, we will explore evidence of other nongenetic inheritance mechanisms that are no less interesting and potentially just as important, but rarely make newspaper headlines. Our aim is not to provide a comprehensive overview of the evidence (which could easily take up an entire book) but, rather, to give the reader a taste of the range of phenomena encompassed by nongenetic inheritance. We hope to show that the "anomalies" outlined in this chapter and the next can no longer be regarded as anomalies at all. Rather, these case studies should be seen as examples of the long-overlooked nongenetic dimension of heredity.

WHAT'S IN A PREFIX?

By now, many people have encountered the word *epi*genetics and some are vaguely aware that this new science somehow undermines the foundations of genetics, like a colony of termites chewing through the beams under a house. The term itself has acquired a kind of mystique—you'll find it on websites devoted to alternative medicine, miracle diets, and "intelligent design." So what is epigenetics, and why is it generating so much excitement? And how does epigenetics fit into the bigger picture of extended heredity?

Part of the confusion surrounding epigenetics is that the meaning of this term has changed over the years, and, today, these varied meanings coexist in the scientific literature, and sometimes on the same page.[109] The term was used in the 1940s by British biologist Conrad Waddington to refer to the interaction between genes and environment that results in the production of a phenotype and, later, became a label for all non-genetic factors that influence the expression of genes (epigenetics in the broad sense). In recent years, however, the term has increasingly been used to refer to the effects of certain types of molecules (in particular, methyl groups, histone proteins, and RNA) that interact with DNA and influence when, where, and how much genes are expressed (epigenetics in the narrow sense).

Adding to the confusion is the fact that the different meanings of epigenetics are usually associated with different research aims. The term *epigenetics* is typically used in the broad sense by evolutionary ecologists and developmental biologists whose goal is to understand the environmental factors that shape phenotypes. In that context, epigenetics is essentially the study of development and plasticity at the whole-organism level. By contrast, epigenetics is used in the narrow sense by molecular biologists seeking to understand the biochemical mechanisms that regulate gene expression. In other words, the broad definition is fundamentally concerned with phenotypic outcomes, whereas the narrow definition is concerned with molecular mechanism. Because the coexistence of these very different definitions introduces unnecessary confusion, we will restrict the term "epigenetics" to its narrow-sense definition in this book.[110]

To complicate matters even further, the narrow-sense definition of epigenetics encompasses at least three distinct types of epigenetic variation, which Eric Richards has labeled "obligatory," "facilitated," and "pure."[111] These distinct types of epigenetic variation differ in their hereditary roles.

Obligatory epigenetic factors are strictly determined by the genomic DNA sequence. Such factors play important roles in regulating the stereotyped progression of events during embryonic development and in determining the features of the different tissue types in our bodies. Within a developing embryo, all somatic cells carry essentially the same DNA sequence. Yet, soon after the zygote begins to divide, different cells in the early embryo acquire distinct epigenetic profiles and then transmit these epigenetic differences to their daughter cells. These epigenetic differences determine where, when, and how much a given gene is expressed, allowing different cell lines to specialize as skin, nerve, muscle, gut, gonad, and other types of tissue. However, the epigenetic profiles of all these tissue types (for example, which gene promoters are heavily methylated and which chromosome regions are tightly bound up with their histone proteins, silencing the genes in those parts of the genome) are themselves encoded in the genome. Different individuals carry different genes, and it's these genetic differences that cause different epigenetic patterns to arise, leading in turn to changes in development and causing different individuals to express different traits. Thus, while obligatory epigenetic factors are important cogs in the machinery of organismal development and functioning, this epigenetic machinery is fully determined by genes and merely forms part of the causal chain of molecular events linking phenotype to genotype. Obligatory epigenetic variation can be regarded as hereditary in the context of cell-to-cell transmission within a multicelled body because, once established, the distinct epigenetic patterns are faithfully transmitted within cell lineages. However, obligatory epigenetic variation plays no hereditary role across generations in multicelled organisms because none of it is transmitted independently of variation in genetic alleles.

By contrast, facilitated and pure epigenetic variation is partly or wholly independent of the DNA sequence. The difference between these two types of epigenetic variation is subtle. Facilitated epigenetic factors are probabilistically related to DNA sequences; for example, certain DNA sequences such as transposons are prone to alternate between methylated or unmethylated states, either spontaneously or in response to specific environmental triggers such as diet or stress. Pure epigenetic variation is entirely independent of DNA sequence variation; it encompasses random epigenetic changes such as those that appear to be involved in the formation of certain cancerous tumors. In practice, it's often very difficult to determine whether a particular epigenetic variant belongs in the facilitated or pure category. However, both facilitated and pure epigenetic variation have the potential to be transmitted across generations independently of genes, resulting in "transgenerational epigenetic inheritance."[112] In other words, just as a genetic mutation (a change in the base-pair sequence of the DNA) can be passed across generations and cause a new phenotype to be expressed in the offspring, so an *epimutation* (that is, an epigenetic change, either spontaneous or induced by an environmental factor) can be passed to offspring, causing the offspring to express a different phenotype. The possibility of epigenetic changes that occur independently of DNA sequence variation and that can be transmitted across generations means that epigenetic variation can contribute to heritable variation.

There is one final complication: transgenerational epigenetic inheritance is only considered to occur when inheritance is actually mediated by the transmission of an epigenetic factor in the egg or sperm (that is, through the germ line). As we will see in chapter 5, many examples of nongenetic inheritance involve epigenetic changes, but these changes are secondary consequences of the transmission of some other factor. For example, the intrauterine environment, or maternal postpartum behavior, can influence offspring development, and the developmental effects can involve epigenetic changes in the brain or body of the offspring. Such effects are not included within the scope of transgenerational epigenetic inheritance because they are not mediated by the transmission of an epigenetic factor through the germ line.

Below, we will outline some putative examples of transgenerational epigenetic inheritance.[113] These examples are grouped by the epigenetic mechanism that has been linked most directly to these effects. However, multiple epigenetic systems are probably involved in all of these examples, and the nature of the epigenetic factor actually transmitted from parents to offspring through the germ line remains unknown in some cases.

INHERITANCE OF DNA METHYLATION

In 1744, the Swedish naturalist Carl Linnaeus, then busy cataloging and systematizing the diversity of earthly life-forms, stumbled upon a

Figure 4.1. Toadflax (*Linaria vulgaris*) in its normal form (*left*) and "monstrous" peloric form (*right*). The peloric form turned out to be an epimutant rather than a genetic mutant. (Illustrations by James Sowerby, John Innes Historical Collections. Courtesy of the John Innes Foundation.)

monster. This was not one of the dragons or satyrs listed under the category "Paradoxa" in his *Systema Naturae*,[114] but a curious variant of the small flowering plant commonly known as toadflax (and later named *Linaria vulgaris* in Linnaeus's honor). Instead of the asymmetrical flowers produced by normal toadflax plants, the "peloric" plants (from the Ancient Greek πέλωρ, "monster") produced flowers with five identical, radially arranged petals (figure 4.1).

Linnaeus's monster lay forgotten in the botany manuals for a quarter of a millennium—until, in 1999, it became the first naturally occurring mutation in any species to be subjected to a detailed DNA-sequence analysis. Expecting to uncover differences between mutant and normal forms of *Linaria* in the base-pair sequence at the *Lcyc* locus, which regulates flower symmetry, researchers were astonished to find that they were essentially identical. Instead of a mutant DNA sequence, the monstrous phenotype was associated with a drastic reduction in DNA methylation of the *Lcyc* gene. DNA methylation (that is, the bonding of methyl groups to some DNA bases, most commonly cytosines) is very widespread across genomes of flowering plants, mammals, and some other organisms and serves to regulate gene expression by interfering with the molecular machinery that translates DNA sequences into RNA templates. Reduced methylation is therefore typically associated with increased levels of gene expression. Moreover, such hypomethylated *Lcyc* alleles could be transmitted from parent plants to their offspring, and transmission of this methylation state evidently accounted for the monster's ability to beget monsters (albeit with lower propagation fidelity than that of the normal form).[115] In other words, Linnaeus's monster was not a genetic mutant but an epimutant.[116] This discovery pointed to the intriguing possibility that some of the heritable phenotypic diversity seen in natural populations is based on differences in DNA methylation patterns rather than DNA sequences, inspiring researchers to start sequencing the "epigenomes" (that is, epigenetic—typically, DNA methylation—patterns across the entire genome) of wild plants and animals.

Even before the epigenetic basis of the toadflax monster was uncovered, researchers observed that DNA methylation affected the expression of a variant of the *Agouti* gene in mice. It was known since the late 1970s that genetically near-identical mice carrying the mutant A^{vy} allele of *Agouti* were extraordinarily variable in both coat color and health, with individuals ranging from dark and healthy to yellow and sickly. Furthermore, maternal phenotype influenced offspring phenotype within isogenic strains, seemingly violating the genotype/phenotype dichotomy. It was eventually discovered that the A^{vy} allele bears a retrotransposon (a "selfish" genetic element that can insert extra copies of itself anywhere in the genome) upstream of its promoter region—the

Figure 4.2. Isogenic mice ranging in phenotype from yellow and sickly on the left to dark and healthy (pseudoagouti) on the right. These phenotypic differences result from variation in epigenetic factors that influence the expression of the *Agouti* gene. (Photo: Jennifer Cropley.)

base-pair sequence that functions as an on/off switch for the gene, and plays a role in determining when and where the gene is expressed. Methylation of this retrotransposon affects the expression of the *Agouti* gene: heavy methylation shuts down the retrotransposon and results in a darkly mottled ("pseudoagouti") coat and good health, whereas reduced methylation of the retrotransposon results in yellow fur and obesity.[117] It was also shown that the availability of methyl-donor substances such as folic acid in the maternal diet affects the methylation state of the A^{vy} allele (figure 4.2). In 1999, Emma Whitelaw's research group in Australia used a series of breeding and egg-transfer experiments to investigate the mechanism linking maternal and offspring phenotypes. They concluded that mothers transmit an epiallele that results in differential methylation of the A^{vy} allele in offspring, providing the first putative example of transgenerational epigenetic inheritance in mammals.[118]

Several years later another Australian research group, led by Catherine Suter, showed that this epigenetically controlled phenotype can respond to natural selection.[119] Jennifer Cropley fed mice a methyl-donor enriched diet, and mated male mice carrying the mutant A^{vy} allele to genetically normal females over five generations. As expected, offspring that inherited the A^{vy} allele from their father ranged in coat color from pseudoagouti to yellow but, in each generation, only the pseudoagouti males were allowed to breed, generating strong selection for the pseudo-agouti phenotype. As a result, over five generations, the proportion of pseudoagouti offspring increased steadily from 29 percent to 49 percent, showing that a purely epigenetic trait can respond to natural selection and produce a shift in the population's mean phenotype. The *Agouti* gene provides an example of an environmentally induced epigenetic variant that has beneficial phenotypic effects, and Cropley's experiment shows how such epigenetic traits could contribute to adaptive evolution.

Yet, Cropley's study also highlighted the difficulty of deciphering the mechanisms involved in epigenetic inheritance. Despite their best efforts, the researchers were unable to elucidate what it was that actually changed at the molecular level to bring about increased pseudoagouti expression. The obvious candidate—the extent of methylation of the A^{vy} allele—did not increase over the five generations of selection, nor was there any change in the abundance of an associated type of RNA. Moreover, although nongenetic transmission of coat color and health in A^{vy} mice is not observed through the father mouse, the methylation of the A^{vy} allele in mouse embryos is only sensitive to maternal diet when this allele is inherited paternally.[120] Even more puzzlingly, the methylation state of the A^{vy} allele appears to be completely erased between generations.[121] The molecular mechanism mediating the transmission of epialleles of the *Agouti* gene therefore remains a mystery.

It also appears that the agouti phenotype is not transmissible beyond the grand-offspring, and some researchers have therefore argued that this example does not unequivocally demonstrate epigenetic inheritance through the germ line (that is, true "transgenerational epigenetic inheritance"). After all, it's possible that substances present in the mother's body could influence not only the embryos in her womb but even the ovules developing inside the female embryos. Thus, only maternal transmission to great-grand-offspring, or paternal transmission to the grand-offspring, can provide conclusive evidence of the transmission of epialleles through the germ line in a mammal.

Such evidence was provided by a discovery made over a decade ago by Matthew Anway, Michael Skinner, and colleagues at Washington State

University. To study the effects of pesticides on embryonic development, they exposed pregnant female rats to chemicals, such as the fungicide vinclozolin, that mimic the sex hormones, upsetting the delicate hormonal balance in the womb. No one was surprised when the male pups born from such mothers suffered poor health and fertility. But, astonishingly, these pups produced sons with similar problems even without reexposure to the chemicals, and so did their sons and grandsons.[122] A close look revealed altered DNA methylation in sperm, suggesting that the low-fertility phenotype is transmitted as an epiallele. More recent work has also identified changes in sperm-borne small RNA.[123]

Other toxins that leach out of plastic items such as toothbrushes, baby bottles, and food-storage containers have since been shown to induce changes in DNA methylation (and perhaps other epigenetic factors as well) in gametes and to affect several generations of descendants, resulting in poor health and behavioral changes.[124] Perhaps the best-studied culprit is bisphenol A (BPA), a compound that leaches out of polycarbonate plastic. Because BPA molecules are similar in their shape and properties to the hormone estrogen, BPA can disrupt subtle chemical signals in the brain and body, and its effects on developing embryos and young children could be especially severe. When BPA was administered to monkeys at levels that are commonly detected in humans, it was found to impede the formation of memories in adults, and, in embryos exposed in utero, to affect the development of reproductive traits and postnatal behaviors.[125] Experiments on mice and rats have also shown that exposure to BPA in utero can lead to obesity and diabetes in the offspring.[126] In mice and fish, BPA exposure has been found to cause reduced fertility in several generations of descendants.[127] The results of such experiments should worry us human inhabitants of a world increasingly polluted with such chemicals—a topic to which we will return in chapter 10.

A recent study suggests that environmentally induced changes in epialleles can be remarkably specific. Brian Dias and Kerry Ressler of Emory University in Atlanta conditioned male mice to associate a particular chemical odor (acetophenone) with a mild electric shock, and then mated these males to naive females. They found that the offspring and grand-offspring of these males were sensitized to acetophenone, responding much more strongly to this chemical by comparison with offspring of control mice sensitized to a different chemical. Both the sensitized males and their descendants exhibited enlarged glomeruli in the brain's olfactory bulb, the neural machinery of odor processing. Paternal transmission of this behavioral and neurological phenotype over at least two generations occurred even when offspring were produced via in vitro fertilization and cross fostered, ruling out the possibility of direct behavioral influence of parents on their offspring. Reduced DNA methylation of the *Olfr151* locus, which codes for the specific olfactory receptor involved in detection of acetophenone, was observed in males sensitized to this chemical as well as in their sons, suggesting that the sensitized phenotype is transmitted to offspring via hypomethylated *Olfr151* alleles. Conditioning of female mice resulted in similar effects on offspring. Dias and Ressler's study therefore shows that mice subjected to a specific sensory experience can transmit altered patterns of gene expression via the germ line, resulting in modified behavior in descendants in response to the same sensory stimulus.[128] This discovery points to the possibility that an individual's responses to smells and tastes may be shaped by hereditary transmission of the learned experiences of ancestors—in other words, that your intense aversion to cocktail prawns could be a nongenetic legacy of a nasty bout of "seafood buffet surprise" suffered by a grandparent who died before you were born.

Importantly, such evidence suggests that some epialleles transmitted by the egg and sperm to the zygote can persist in the embryo and influence its development. Each tissue in the body is characterized by a specific epigenetic signature that induces a specific pattern of gene expression and gives the tissue its distinctive features. However, the zygote genome must begin with a relatively clean slate in order to have the capacity to give rise to all tissue types (totipotency). To make this possible, the primordial germ cells developing inside the embryo undergo extensive epigenetic reprogramming to bring them to an "epigenomic basal state."[129] When the embryo matures into an adult, mates, and its germ cells participate in the formation of a zygote, that zygote will thus be able to produce all the tissues that make up the adult body, with their distinctive epigenetic profiles. However, the epigenomes of bodily cells, including germ-line cells, also undergo epigenetic changes over the course of an individual's lifetime, and inheritance of environmentally induced epialleles requires that some of these acquired epigenetic changes can be transmitted to the next generation. In mammals, the maternally inherited chromosomes retain much of their epigenetic profile in the zygote, whereas the paternally inherited chromosomes undergo extensive DNA demethylation and chromatin remodeling.[130] This suggests that there is more scope for transgenerational epigenetic inheritance through the egg than through the sperm in mammals. However, the discoveries described here suggest that some acquired epigenetic marks in the sperm can survive epigenetic reprogramming and be transmitted to the embryo as well. Intriguingly, the opposite pattern is observed in some organisms. For example, in the tiny zebrafish, the methylation state of paternally inherited genes persists in the embryo while the methylation state of maternally inherited genes is converted to the paternally inherited pattern,[131] suggesting that environment-induced changes in methylation of sperm-borne DNA could be a key mechanism of nongenetic inheritance in this species.

The studies outlined here suggest that transmission of differentially methylated DNA regions from mothers and fathers to their offspring could be an important mechanism of nongenetic inheritance, and that the sensitivity of DNA methylation to environment could allow for the transmission of environmental effects to descendants. However, many questions remain. The role of DNA methylation in regulating gene expression has been most extensively studied in plants, and the picture emerging from that work is mixed. Although genetic variation appears to be more important overall, variation in DNA methylation plays an independent role in regulating some genes, and this role could be substantially underestimated by laboratory studies that eliminate most environmental variation.[132] Compounding these uncertainties about the independent role of DNA methylation within generations is uncertainty about the potential for transmission of DNA methylation patterns across generations. It's still unclear how and to what extent the DNA methylation patterns transmitted in gametes can be retained in the developing embryo given that methylation is extensively reprogrammed after fertilization, and it is even less clear how DNA methylation variants could persist through multiple generations.[133] Moreover, while the genomes of some organisms (like vertebrates and flowering plants) are rife with DNA methylation, this epigenetic system appears to play a lesser role in other organisms (like many insects). While the role of DNA methylation in heredity continues to be explored, some researchers are turning their attention to another epigenetic mechanism—transmission of RNA from parents to their offspring.

INHERITANCE OF RNA

The earliest stages of development occur before the embryo's genome is "activated" and begins to be expressed, meaning that the first cycles of cell division (and the complex changes that occur in these cells, with potentially lifelong consequences) occur while the embryo is fully under the control of factors supplied by its parents. In other words, the early embryo operates on a parentally programmed autopilot whose settings are determined by the nongenetic components of the egg and sperm cell, including the large cargos of RNA that they contribute to the zygote. The RNAs transferred in the gametes are transcripts of the parental genomes, not the offspring genome, and the nature and quantity of parental RNA has the potential to influence the course of offspring development. Eggs are packed with maternal RNA, and recent evidence suggests that sperm also deliver a rich complement of RNA molecules.

Some RNAs transmitted by a parent to its offspring are ultimately diluted by RNAs transcribed from the offspring's own genome, such that any RNA-mediated hereditary effects fade out over one or two generations. However, in some cases, RNA transmitted in sperm or egg can somehow regenerate itself in the offspring, allowing effects of ancestral environment to be transmitted for multiple generations. Intriguingly, work on nematode worms suggests that such self-regenerating RNA inheritance could be highly regulated, with some cellular systems working to bring about the self-regeneration of the RNA and other systems functioning as off switches that terminate the transmission of an environmental effect after a certain number of generations. Such regulation of the duration of transmission is to be expected if RNA inheritance functions as an adaptive mechanism that allows parents to hone the features of their offspring for the environment that the offspring are likely to encounter.[134]

The variety of RNAs transmitted in the gametes is bewildering, and their roles in development are still poorly understood. There are

PIWI-interacting RNAs (piRNA), micro-RNAs (miRNA), and transfer RNA-derived RNAs (tsRNA), together affectionately referred to as "small noncoding RNA" (sncRNA) to differentiate them from long noncoding RNAs (lncRNA). Messenger RNA is present in both eggs and sperm, and eggs also contain ribosomal RNA (rRNA) that makes up part of the structure of ribosomes—the organelles that translate mRNA transcripts into protein. RNAs can also be differentially methylated, resulting in different three-dimensional structure and probably affecting their biological function.[135] These RNAs can influence embryonic development in a number of ways. Noncoding RNAs are, by definition, not translated into protein. Instead, they are thought to have regulatory roles, influencing which genes are expressed when, where, and how much. For example, some sncRNAs appear to act by binding to mRNAs and thereby disrupting their translation into peptides—a mechanism known as "RNA interference" (RNAi). lncRNAs may bind to the chromosomes, alter their three-dimensional structure (known as "chromatin structure"), and thereby alter the expression of genes in the affected genomic regions.[136]

While many of these RNAs are transcribed from parental genes during the process of gamete formation in the gonads, recent research has revealed an intriguing twist to this story. It appears that cells throughout the body can secrete tiny membrane-enclosed bubbles of cytoplasm into the blood plasma and other bodily fluids, and these minute "extracellular vesicles" can travel through the body and fuse with other cells, including gametes. Vesicles are packed with RNA and other molecules, and vesicles carrying a range of RNA types have been detected in human semen. These tiny intercellular parcels appear to be able to adhere to or fuse with sperm cells and thus hitch a ride into the egg.[137] The bodily fluids also contain free RNA molecules, often bundled together with proteins. Intriguingly, studies on nematode worms even show that RNA secreted by a worm's neurons (of which a typical worm has exactly 302) can find its way to other bodily tissues and that, when such brain-derived RNA enters the germ line, it can be transmitted to offspring and alter the expression of their genes.[138] Although the role of extracellular vesicles and free RNA in heredity and development remains poorly understood, a number of researchers have noted the uncanny resemblance of these particles to the "gemmules" that Darwin postulated in his long-derided theory of pangenesis. Indeed, if vesicles secreted by various bodily tissues can travel to the gonads and find their way into gametes, these microscopic parcels of information could mediate the transmission of environmental influences across generations, just as Darwin had supposed.[139]

The first compelling evidence of RNA-mediated inheritance in animals was furnished by Minoo Rassoulzadegan and coworkers in France over a decade ago.[140] The researchers noticed that mice heterozygous for an artificially engineered mutation of the *Kit* gene produced offspring that often expressed the mutant phenotype—a white tip on the tail—even if they did not inherit the mutant allele, and these genetically normal offspring also tended to transmit the white tail phenotype to their own offspring. Such puzzling hereditary patterns, where a genetic mutation in an ancestor somehow alters the expression of normal genes in its descendants, had been described earlier in plants, and dubbed "paramutation." Rassoulzadegan and coworkers found that the sperm of mutant mice contained abnormal RNA, and suspected that this RNA was the factor responsible for paramutation in offspring. To test this conjecture, they obtained the mutant RNA and injected it into genetically normal, fertilized mouse eggs (zygotes). These zygotes developed into adult mice that not only expressed the mutant phenotype themselves but also passed this phenotype to their offspring. This shows that mutant RNA that finds its way into the gametes can "paramutate" genetically normal offspring, causing them to express the mutant phenotype as well as produce and pass on the abnormal RNA to their own offspring.

White tail tips may seem harmless enough, but a subsequent study by Rassoulzadegan's research group also demonstrated the potential for paramutation of the *Cdk9* gene, which controls heart development. Kay Wagner and coworkers injected a micro-RNA that regulates the expression of *Cdk9* into fertilized mouse eggs, and found that these eggs developed into mice with severely enlarged hearts. Like white tail tips, this heart abnormality was also transmitted to the genetically normal offspring of paramutated mice via sperm-borne RNA.[141]

Although the quantity of RNA in sperm is miniscule by comparison to that of the egg, there is mounting evidence of an important role for sperm-borne RNA, and several studies have found evidence of transmission of environmental effects to offspring via this mechanism. In one study, researchers subjected male mice to the stress of repeated separation from their mother in infancy and found that the offspring and grand-offspring of these traumatized males had altered stress responses. Moreover, when the researchers extracted and purified RNA from the sperm of these males and injected this RNA into normal zygotes, the resulting offspring expressed similar stress responses, and even passed this stress phenotype to their own offspring. The researchers were able to identify nine sperm-borne miRNAs as the causal factors transmitted from traumatized males to their offspring, and showed that these miRNAs appeared to act by binding to and disabling the translation of maternally derived mRNA in the early embryo.[142] In other words, nongenetic factors transmitted in the sperm of stressed males appear to interfere with maternally derived nongenetic components and thereby reprogram the embryo's autopilot, setting it on a course to develop into an adult with depression-like symptoms and altered responses to stressful situations.

In another study, RNA extracted from the sperm of males raised on a high-fat diet and injected into normal zygotes produced offspring that suffered from some of the same metabolic symptoms as their obese, glucose-intolerant father. In this case, transmission of the symptoms from father to offspring appeared to be mediated by sperm-borne tsRNA, and the authors proposed that these small RNAs might bind to the promoter sites of genes involved in glucose metabolism and thereby alter the expression of those genes.[143] These findings raise the possibility that paramutation may be involved in the transmission of hereditary diseases in our own species. After all, human sperm carries a rich RNA cargo whose functions in embryonic development are largely unknown.[144]

RNA inheritance can also mediate the transmission of environmentally induced effects across generations in plants, and a recent study in the common dandelion suggests that such effects could extend beyond a single generation. Dandelions exposed to two environmental stressors—water deprivation, and exposure to the chemical salicylic acid—transmitted an altered complement of small RNAs to their offspring, and those offspring (which were not themselves exposed to the stressors) also transferred altered small RNA cargos to their own offspring. Ancestral exposure to stressful environments thereby affected the expression of a number of genes in offspring and grand-offspring.[145]

INHERITANCE OF CHROMATIN STRUCTURE

The least well-understood type of epigenetic inheritance involves the transmission of chromatin structure—that is, the three-dimensional structure of chromosome regions in eukaryotic cells. Chromatin structure is largely determined by the way in which the DNA molecule is wrapped around special proteins called histones to form bundles called nucleosomes. Like the transmission of DNA methylation and RNA, the transmission of chromatin structure in gametes can influence the expression of genes in the embryo. In regions of chromosomes that are less tightly bound up, the DNA is more exposed to the molecules (called "transcription factors") that initiate the transcription of genes into RNA templates, and gene expression therefore tends to occur at a higher rate. Conversely, the DNA in tightly bundled regions is shielded from transcription factors, so gene expression is largely repressed. Interestingly, just as DNA methylation affects gene expression, so the structure of nucleosomes and consequent rate of gene expression depends in part on the binding of methyl and acetyl chemical groups to the histones. Chromatin structure is highly dynamic and sensitive to environment, meaning that the transmission of environment-induced variations in chromatin structure from parents to offspring could mediate epigenetic inheritance.[146] Indeed, maternal DNA in the egg comes bundled into nucleosomes that could retain a memory of environments experienced by the mother. However, the potential for paternal transmission of chromatin structure is less clear because sperm-borne DNA is largely free of histones, and its chromatin structure is mostly (although not completely) reset in the zygote through the addition of histones supplied in the egg.[147]

Evidence of epigenetic inheritance via the transmission of chromatin structure is beginning to accumulate. In 2003, Vincent Sollars and colleagues showed that abnormal outgrowths from the compound eyes of *Drosophila melanogaster* occurred as a result of reduced histone acetylation at certain genomic regions. The variable eye morphology was observed in nearly isogenic lines and appeared to be inherited independently of any genetic alleles. When they selected for this abnormal eye morphology, they observed a rapid increase that was stable while selection was maintained but was rapidly reversed under selection for normal eyes.[148] More recently, it was shown that *Drosophila* males fed a high-sugar diet sire offspring that are prone to overeating and obesity. The authors observed changes in chromatin structure in the sperm of sugar-fed males as well as their offspring and showed that these changes were associated with altered gene expression in certain genomic regions.[149]

Even more impressive results have been obtained with the tiny nematode worm *Caenorhabditis elegans*. When RNA interference (RNAi) was used to silence the worm's genes, it was found that this effect could persist for many generations—indeed, perhaps indefinitely—via transmission of altered chromatin structure. For example, feeding worms RNAi targeting the *ceh-13* gene resulted in a phenotype that the researchers described with technical flair as "small and dumpy," and when such worms were selected for breeding in each generation, some of their descendants were born small and dumpy as well. Moreover, transmission of dumpiness was not dependent on the action of genes that encode proteins essential for the RNAi silencing mechanism, but genes that regulate chromatin structure were essential for transmission to occur.[150] RNAi-induced silencing of a number of other genes was shown to be transmitted in a similar way. This study therefore showed that a single episode of gene silencing by RNAi could induce a phenotype that could be stably transmitted over many generations when favored by selection, suggesting that such nongenetic traits could play a role in adaptation. More recently, it was shown that lifespan is strongly dependent on histone methylation in *C. elegans*, and that these patterns of histone methylation are transmitted to offspring and grand-offspring.[151]

Importantly, new evidence also suggests that highly stable chromatin changes can be induced by the environment. Adam Klosin and colleagues reported that a change in temperature can alter patterns of gene expression in *C. elegans*, and these acquired traits can then be transmitted through both eggs and sperm over at least fourteen generations. These effects appear to be mediated by temperature-induced changes in methylation of histone proteins associated with certain genes.[152]

Evidence that such epigenetic traits can be induced by environmental triggers links these effects with the ecology of natural populations. For example, if a change in gene expression induced by high temperature provides a fitness advantage, then the frequency of this epigenetic variant could be increased by natural selection, and this trait could potentially play a role in adaptation to elevated thermal stress.

The case studies that we have outlined in this chapter provide some compelling evidence for the occurrence of epigenetic inheritance through the germ line in multicelled organisms (that is, "transgenerational epigenetic inheritance"), but this is not the only form of nongenetic inheritance. Many other nongenetic mechanisms that allow for the transmission of information from parent to offspring are known, and these too can have important evolutionary and medical implications. In the next chapter, we will explore these less widely appreciated forms of nongenetic inheritance.

5

The Nongenetic Inheritance Spectrum

> The more we learn about the world, and the deeper our learning, the more conscious, specific and articulate will be our knowledge of what we do not know, our knowledge of our ignorance.
>
> —Karl Popper, *Conjectures and Refutations*, 1969

In the excitement surrounding epigenetic discoveries, it's easy to forget that epigenetic inheritance in the narrow sense comprises only a few among many mechanisms of nongenetic inheritance. Epigenetic inheritance is fascinating, widespread, and important, but the scope of nongenetic inheritance is much greater. Moreover, while narrow-sense epigenetics is a relatively young research field, the scientific literature on nongenetic inheritance actually stretches back a century, and many interesting studies are now all but forgotten. Although some of the phenomena described in this literature may turn out to be examples of transgenerational epigenetic inheritance, it's clear that many examples of nongenetic inheritance involve other mechanisms of transmission. Many of these examples are encompassed by the broad category of phenomena known as maternal and paternal effects (or, collectively, parental effects). Although parental effects have been recognized for many years and are known to have important functions in many species, they have seldom been incorporated into evolutionary models, and their general role in evolution remains poorly understood.[153] We believe that these diverse effects can be understood as instances of nongenetic inheritance and can be included, along with epigenetic inheritance, within the scope of extended heredity.

In this chapter, we will outline a range of examples of nongenetic inheritance that appear to fall outside the bounds of transgenerational epigenetic inheritance in the strict sense[154] and consider the ecological and evolutionary roles of such effects. In these examples, inheritance does not appear to be mediated by the transmission of an epigenetic factor (that is, a DNA methylation pattern, RNA, or chromatin structure) through the germ line. Many of these examples probably involve epigenetic changes such as altered patterns of DNA methylation in the body of the offspring during development or throughout life, but these epigenetic changes are a downstream consequence of the transmission of some other type of factor at conception or via subsequent parent-offspring interaction. We will then briefly explore the challenges involved in detecting nongenetic effects and identifying the factors and processes mediating their transmission. Having surveyed diverse examples of nongenetic inheritance, in the final section of this chapter we will take a step back and ask why nongenetic inheritance exists at all.

PARENTAL LEGACIES

One category of nongenetic effects—maternal effects—are so obvious that their existence has been recognized for several decades. By definition, maternal effects occur when maternal phenotype affects offspring phenotype, and this effect cannot be explained by the transmission of maternal alleles.[155] Such effects can occur via the myriad routes of influence that mothers have on their offspring, including transgenerational epigenetic inheritance (some examples of which were outlined in the previous chapter), variation in the structure of the egg cell, the intrauterine environment, the mother's choice of the location where the eggs or offspring are brought into the world and modification of the environment that the offspring will experience, as well as postpartum physiological and behavioral interactions (figure 5.1). Some maternal effects are passive consequences of maternal features for offspring development (including deleterious effects of maternal poisoning, illness, or senescence), while other maternal effects represent reproductive investment strategies that have evolved to enhance reproductive success.[156] Such effects can therefore either enhance or reduce the fitness of mothers and their offspring.

Figure 5.1. In many species, mothers care for their offspring after birth. *Lyramorpha rosea* bug nymphs (*left*) are guarded and perhaps fed by their mother. A baby gorilla (*below*) spends eight months in its mother's womb and then several years in its mother's care, drinking its mother's milk and learning how to be a gorilla. Such mother-offspring interactions create many channels for postfertilization nongenetic transmission of maternal influence to offspring. In some species, fathers care for their offspring and have analogous opportunities to influence offspring development nongenetically. (Photos: R. Bonduriansky)

Until quite recently,[157] maternal effects were regarded as little more than a nuisance—a source of environmental "error" in genetic studies. But geneticists at least felt assured that, in most species (including key laboratory "model organisms" such as flies and mice), fathers could transmit nothing more than genetic alleles to their offspring.[158] Yet, recent research has revealed numerous examples of paternal effects in mice, *Drosophila*, and many other species as well.[159] In fact, in sexually reproducing species, paternal effects could turn out to be as widespread as maternal effects.

Offspring can be influenced by parental environment and experience, parental age, and parental genotype. An environmental factor (such as a toxin or a nutrient) could induce changes in the parental body that affect offspring development. As we will see, the deterioration of the body with age can also affect reproductive traits and heritable nongenetic factors, and thereby affect offspring development.

Cases where a gene expressed in a parent influences the phenotype of its offspring are known as "indirect genetic effects."[160] Perhaps counterintuitively, such effects fit within the scope of nongenetic inheritance because they are mediated by the transmission of nongenetic factors. For example, a particular gene expressed in a parent might affect its behavior toward its offspring, or alter the epigenetic profile of other genes in the germ line, and thereby affect the development of offspring even if they do not inherit that gene. A striking example of an indirect genetic effect was provided by a study on mice. Vicki Nelson and colleagues crossed different inbred mouse strains to produce males that were genetically similar in every way except for their Y chromosome. They then asked a very odd question: does a male's Y chromosome affect the phenotype of his daughters? Anyone who stayed awake through their high-school biology classes knows that daughters don't inherit their father's Y chromosome, so, by the logic of classical genetics, genes on the paternal Y chromosome cannot affect daughters. But Nelson and colleagues found that the identity of the paternal Y chromosome did affect a variety of physiological and behavioral traits in daughters. In fact, the effects of the paternal Y chromosome on daughters were comparable in magnitude to the effects of the paternal autosomes or X chromosome, which daughters do inherit. Although the mechanism involved remains unknown, genes on the Y chromosome must have somehow altered the sperm cytoplasm, sperm epigenome, or the composition of the seminal fluid, allowing genes on the Y chromosome to influence the development of offspring that did not inherit those genes.[161]

PRIMED FOR SUCCESS OR DOOMED TO FAILURE?

Some maternal and paternal effects appear to have evolved as a means to give offspring a head start in the type of environment that the offspring are likely to encounter.[162] A classic example of such "anticipatory" parental effects is the induction of defenses in offspring of parents exposed to predators. "Water fleas" (*Daphnia* sp.) are tiny freshwater crustaceans that swim with a slow, jerking movement, using a pair of long appendages as oars. They are easy prey for predaceous insects, crustaceans, and fish. When they encounter chemical cues from predators, some species of *Daphnia* respond by growing spines on their head and tail, thus making themselves harder to grab or swallow (figure 5.2). *Daphnia* exposed to predator cues also produce offspring that develop spines even in the absence of predator cues and may undergo changes in growth rate and life history that further reduce vulnerability to predation. Such transgenerational induction of antipredator defenses also occurs in many plants; when attacked by herbivores such as caterpillars, plants produce seedlings that secrete unpalatable defensive chemicals (or are primed to initiate such defenses more rapidly in response to herbivore cues), and such induced defenses can persist for several generations.[163]

While it's not yet known how *Daphnia* mothers induce spine development in their offspring, some examples of apparently adaptive maternal and paternal effects involve the transfer of particular compounds to offspring. For example, *Utetheisa ornatrix* moths obtain pyrrolizidine alkaloids by eating legumes that synthesize this toxin. Females are attracted to the scent of males that are well-endowed with this chemical, and such males transfer part of their toxic store as a "nuptial gift" in the seminal fluid. Females incorporate these alkaloids into their eggs, making their offspring unpalatable to predators.[164]

Parents can also prime their offspring for the social environment and lifestyle that they are likely to encounter, as illustrated by the

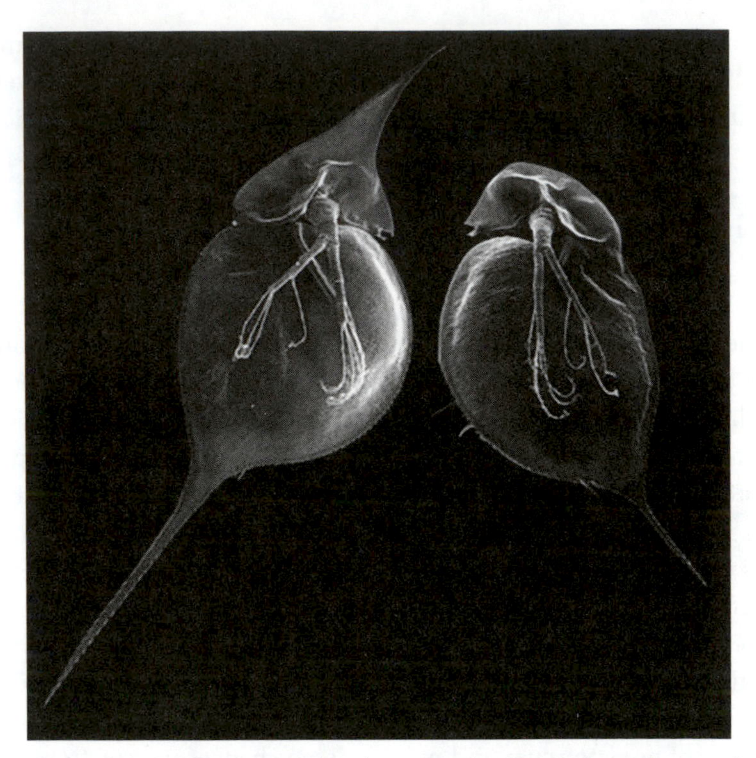

Figure 5.2. In some *Daphnia* species, females exposed to chemical cues from predators develop defensive spines. These females also produce offspring that develop spines even in the absence of predator cues. (© R. Tollrian and C. Laforsch)

desert locust (*Schistocerca gregaria*). These insects can switch between two strikingly different phenotypic morphs: a green-gray "solitarious" morph, and a black-yellow "gregarious" morph. Gregarious locusts are characterized by lower fecundity, shorter lifespan, a larger brain, and a tendency to aggregate into enormous migratory swarms that can denude entire regions of vegetation. Locusts switch rapidly from solitary to gregarious behavior upon contact with a dense locust aggregation, and the population density experienced by females prior to mating also determines the morph of their offspring. Interestingly, however, the full suite of phenotypic changes builds up over several generations, indicating that this maternal effect is cumulative. The effect appears to be mediated by substances transmitted to the offspring via the egg cytoplasm and/or via accessory gland products that coat the eggs, although epigenetic modifications in the germ line could also play a role.[165]

However, parental experience does not necessarily prime offspring to perform better. For one thing, parents could misinterpret environmental cues, or the environment could change too rapidly, and this means that parents will sometimes adjust offspring traits for the wrong conditions. For example, if *Daphnia* mothers induce spine development in their offspring but predators then fail to materialize, the offspring will end up paying the costs of developing and bearing spines but reap no benefits from this trait. In such cases, anticipatory parental effects could actually harm offspring.[166] More generally, offspring face the complex problem of integrating environmental cues received from their parents with cues received directly from their own environment, and their best developmental strategy will depend on which set of cues happens to be more useful and reliable.[167]

Anticipatory effects can misfire but, on the whole, such effects are still expected to be favoured by natural selection. However, many parental effects are not adaptive in any sense at all. Stress may have deleterious effects not only on the individuals that experience it but also on their descendants. For example, research by Katie McGhee, Alison Bell, and colleagues at the University of Illinois showed that female sticklebacks that were exposed to mock predator attacks produced offspring that were slow learners, failed to behave appropriately when confronted with real predators, and were thus more likely to be eaten than offspring of naive mothers.[168] These effects are reminiscent of the dire consequences of maternal smoking during pregnancy in our own species. Correlational studies on human cohorts (and experimental studies on rodents) show that, rather than priming the developing fetus for resistance to respiratory challenges, maternal smoking appears to alter the intrauterine environment in ways that predispose the child to reduced lung function and asthma, in addition to reduced birth weight, psychological disorders, and other problems.[169] Similarly, in organisms ranging from yeast to humans, old parents tend to produce sickly or short-lived offspring.[170] Although the transmission of genetic mutations via the germ line could contribute to such "parental age effects," nongenetic inheritance appears to play a major role (we will consider parental age effects in more detail in chapter 9). Thus, although some types of parental effects represent evolved mechanisms that can enhance fitness, it is clear that some parental effects transmit pathology or stress. Such nonadaptive parental effects are comparable to deleterious genetic mutations, although they differ from genetic mutations in being consistently induced under particular conditions.

The fact that parental effects can sometimes be deleterious suggests that offspring should evolve ways to mitigate the harm, perhaps by blocking out certain types of nongenetic information received from their parents. This can occur even if the fitness interests of parents and their offspring are well aligned, since the transmission of erroneous environmental cues or parental pathology will be disadvantageous for both parents and their offspring. However, as Dustin Marshall, Tobias Uller, and others have noted, the fitness interests of parents and their offspring are rarely identical, and parental effects could therefore sometimes become an arena for parent-offspring conflict.[171] Individuals are selected to allocate their resources in a way that maximizes their own fitness.[172] Whenever an individual can expect to produce more than one offspring in its lifetime, it therefore faces a decision about how to carve up the pie among multiple progeny. For example, mothers might maximize their reproductive success by producing a greater number of offspring, even if this means investing less in each offspring.[173] But, since each individual offspring could benefit by getting more resources from its mother, such "selfish" maternal strategies will be costly for offspring, and might select for counterstrategies that enable offspring to extract more resources from their mother.

To complicate matters even further, the interests of the mother and father might diverge as well. For example, as David Haig has pointed out, fathers might often benefit by helping their offspring to extract extra resources from their mother, even if this extra investment reduces the mother's fitness. This is because, whenever males have an opportunity to sire offspring with multiple females, each of whom is also likely to mate with other males, a male's best strategy is to selfishly exploit the resources of each of his mates for the benefit of his own offspring.[174] Such parent-offspring and mother-father conflicts over parental investment are a potentially important but little-explored dimension of the evolution of nongenetic inheritance.

YOU ARE WHAT YOUR ANCESTORS ATE

Of all the countless factors that compose an animal's environment, diet stands out as particularly important for Darwinian fitness, health, and

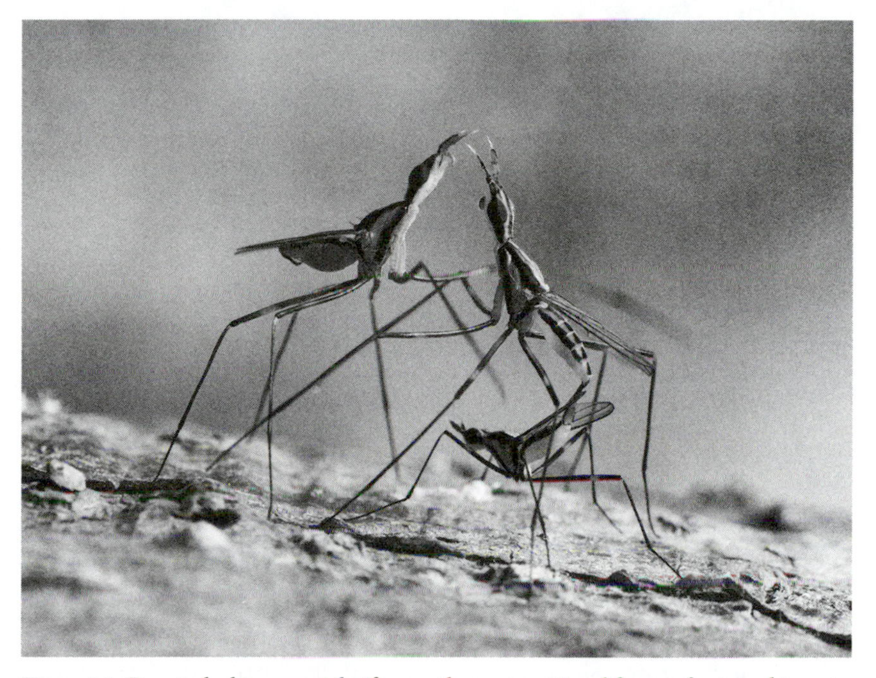

Figure 5.3. Despite lacking nuptial gifts or other conventional forms of paternal investment, males of the neriid fly *Telostylinus angusticollis* that are provided with abundant nutrients as larvae produce larger offspring. Here, two males are seen fighting over a female, who is mating with the male on the right. (Photo: R. Bonduriansky)

many other traits. Perhaps not surprisingly, it turns out that diet can also have major effects across generations. One of us has been studying the effects of diet in the beautiful neriid fly *Telostylinus angusticollis*, which breeds on rotting tree bark along the east coast of Australia (figure 5.3). Neriid males are remarkably variable: in a typical aggregation on a tree trunk, one can spot 2-centimeter-long monsters alongside 5-millimeter-long dwarfs. Yet, when the flies are reared on a standard larval diet in the lab, all adult males turn out rather similar in size, indicating that much of the variation seen in the wild is environmental rather than genetic in origin; in other words, maggots lucky enough to encounter a nutrient-rich food patch develop into large adults, while those relegated to a nutrient-poor patch end up small.

But is any of this enormous environmentally induced variation in male phenotype transmitted across generations? To find out, we generated variation in male body size by rearing some larvae on a nutrient-rich larval medium, while rearing their siblings on a diluted medium.

This resulted in sets of large and small brothers, which we then paired with females that had all been reared on the same larval food. When we measured the offspring, we found that large males produced larger offspring than their small brothers, and subsequent work showed that this nongenetic paternal effect is probably mediated by substances transferred in the seminal fluid.[175] However, because *T. angusticollis* males transfer a tiny ejaculate, orders of magnitude smaller than the typical nutrient-laden ejaculates produced by males of some insect species, this effect does not appear to involve the transfer of nutrients from males to females or to their offspring.[176]

We have recently discovered that such effects can even extend to offspring sired by other males.[177] Angela Crean produced large and small males as described earlier, and mated each female to both types of males. The first mating occurred while the female's eggs were immature, but the second mating occurred two weeks later, after the eggs had matured and become encased in an impermeable shell. Females laid eggs soon after this second mating, and the offspring were collected and genotyped to determine paternity. Since fly eggs can only be fertilized when mature (with the sperm entering through a specialized opening in the eggshell), and females rarely store sperm for as long as two weeks, we were not surprised to see that almost all of the offspring had been sired by the males that mated second. Intriguingly, however, we found that the body size of the offspring was influenced by the larval diet of their mother's first mate. In other words, offspring were larger when their mother first mated with a male that had been well fed as a maggot, even if this male was not their father. A separate experiment ruled out the possibility that females were adjusting their investment in eggs based on their visual or pheromonal assessment of the first male, leading us to conclude that molecules from the first male's seminal fluid were absorbed by the female's immature eggs (or, alternatively, somehow induced females to alter their investment in the developing eggs), and thereby affected the development of embryos sired by a different male. Such nonparental transgenerational effects (dubbed "telegony" by August Weismann) were widely discussed in the scientific literature before the advent of Mendelian genetics, but the early evidence was less than convincing. Our study provided the first modern confirmation that such effects are possible.[178] Although telegony falls outside the scope of heredity in the usual sense of "vertical" (parent-offspring) transmission, it strikingly illustrates the potential for nongenetic inheritance to violate Mendelian assumptions.

There is ample evidence that parental diet can affect offspring in mammals as well. Experimental research on the effects of diet in rats—particularly the restriction of key nutrients such as protein—began in the first half of the twentieth century with the objective of gaining insight into the health consequences of malnutrition. In the 1960s, researchers were intrigued to discover that female rats fed a low-protein diet during pregnancy produced offspring and grand-offspring that were sickly and scrawny and had relatively small brains with a reduced number of neurons, scoring poorly on tests of intelligence and memory.[179] In recent years, research efforts have turned to understanding the effects of excessive or unbalanced nutrient intake, using rats and mice as experimental models to gain insight into the human obesity epidemic, and it is now well-established that both maternal and paternal diets can have a variety of effects on offspring development and health. Some of these effects come about via epigenetic reprogramming of embryonic stem cells in the womb. For example, in rats, a high-fat maternal diet has been shown to reduce the proliferation of hematopoietic stem cells (which give rise to blood cells), while a maternal diet rich in methyl-donors has been found to promote the proliferation of neuronal stem cells in embryos.[180] In rats, a high-fat paternal diet has been found to cause reduced insulin secretion and glucose tolerance in daughters.[181] As we will see in chapter 10, there is evidence of such effects in humans as well.

THE LASTING EFFECTS OF PARENTAL CARE

Parental behavior can have a profound influence on brain development and behavior of offspring. In mice, the stress of early separation from mother results in epigenetic changes in the brain as well as emotional and cognitive impairment. A recent study showed that, when female mice that had been subjected to such early separation stress as pups produced their own offspring, those offspring exhibited similar signs of stress in their brains and behavior. The researchers concluded that the effects of maternal stress were transmitted via changes in maternal care behavior toward the pups, because pups cross fostered from birth between stressed and nonstressed mothers resembled their foster mother.[182] There is evidence that maternal care can influence offspring personality in birds as well. For example, researchers found that foster chicks raised by inexperienced mothers reacted more fearfully to novel experiences than chicks raised by experienced mothers.[183] Cross-fostering studies like these can differentiate the effects of maternal behavior on offspring development from genetic, epigenetic, and intrauterine effects. Correlational evidence from large-cohort studies on human populations suggests that, as in mice and birds, a human mother's behavior toward her child influences its developing brain, and such influence can have long-term consequences for personality and behavior.[184]

Although long ignored, the possibility that fathers' behavior can also influence offspring development has now been confirmed as well. In species lacking paternal care, such paternal effects can operate indirectly, via male effects on the resource investment or care provided by the mother. For example, in mice, social isolation results in increased anxiety. A recent study from the lab of Frances Champagne at Columbia University showed that female mice mated to socially isolated males exhibited reduced maternal care behaviors such as licking and nursing that, in turn, had a negative effect on the growth rate of their pups.[185] In contrast, when fathers interact with their offspring, variation in paternal care behavior can influence offspring development directly. Although paternal care is unusual in rodents, California mouse (*Peromyscus californicus*) fathers participate in rearing their pups. Researchers manipulated paternal care behavior by removing the mother and placing the pups outside the nest, thereby motivating the father to retrieve them. They also castrated some males, which has the effect of reducing paternal grooming behavior. They found that pups subject to increased rates of paternal retrieval were more assertive in territorial defense as adults, and that paternal grooming behavior influenced pup brain chemistry.[186] Interestingly, the potential for such paternal care effects appears to extend well beyond cuddly mammals. In three-spined sticklebacks (*Gasterosteus aculeatus*), mothers provision their offspring with yolk, but fathers bear the full burden of subsequent care, defending their babies from predators, fanning them with their fins, and retrieving them when they wander away from the nest.

When hatchlings were experimentally deprived of paternal care, they matured into fish that were more anxious and, consequently, more vulnerable to predators.[187]

SOCIAL LEARNING, BEHAVIORAL TRADITIONS, AND CULTURE

Our species is unique in the scope and fidelity of cultural transmission and the capacity of every generation to preserve its cultural heritage and to build on it through innovation. It is this capacity for teaching and learning that enables human populations to undergo seemingly limitless, cumulative cultural evolution, making possible the development of complex languages, folklore, technology, and science.[188] Our individual capacity for innovation is limited, but collectively, over many generations, we are able to transform ourselves and the world around us almost beyond recognition. Indeed, Kim Sterelny has argued that the evolution of this unique capacity to learn from our elders and teach our youngsters was the pivotal factor in the history of the hominin lineage that enabled the evolution of our large brains and high intelligence.[189] Behavioral variation can be transferred between unrelated individuals (that is, "horizontally"), but there is no doubt that parent-offspring transmission plays a key role in the preservation and spread of behavioral variation and cultural traditions.

The capacity to transmit behavioral variation and develop culture-like traditions is present in other social and cognitively complex animals such as apes, monkeys, dolphins, and birds, albeit on a more modest scale. The key difference between humans and other animals is in the potential for cumulative cultural evolution—that is, the potential for a cultural innovation to be combined with subsequent innovations to create more complex cognitive or technological tools. This capacity is very limited in other animals. Without it, the nut-cracking stone will never become a hand ax, and the stick will never become a stone-tipped spear.

Yet, even simple cultures can play important ecological roles. In wild chimpanzees, the manufacture and use of rudimentary wood and stone tools is an essential part of life: tools are used to "fish" for termites, crack open nuts, kill prey, intimidate predators and competitors, and in many other tasks. Moreover, chimpanzee populations vary in their tool kits, indicating the potential for cultural variation and adaptation to local conditions.[190] The complexity of some chimpanzee foraging techniques (which may involve prior planning and multiple types of tools) suggests that chimpanzees are even capable of a limited form of cumulative cultural innovation.[191] Most intriguingly, chimpanzees have the capacity to adopt arbitrary cultural conventions, as shown by a group of chimpanzees in Zambia that spontaneously adopted the habit of inserting a blade of grass into one ear.[192] This practice was observed in most members of the group and continued even after the inventor of the behavior had died, but was almost never observed in neighboring groups. Such arbitrary "fashions" could serve as badges of group membership in chimp societies, much as they do in humans.

The potential for behavioral innovation, and the consequences that such innovations can have on a group's ecological strategy when they are transmitted via social learning (that is, learning by observing and imitating other individuals), was famously illustrated by Japanese macaques.[193] Researchers interested in animal behavior were once satisfied to observe pigeons, rats, and even monkeys in cages, but, beginning in the 1940s, a team of Japanese primatologists devised the revolutionary technique of observing wild animals—a *Macaca fuscata* troop on Koshima islet—in their natural habitat while keeping track of each individual and its biological and social relationships within its group. In order to observe interactions around feeding, the primatologists began to provide sweet potatoes to the monkeys, and soon noticed a young female taking her potatoes to a nearby river, dipping them in the water and brushing the sand off with her hand. Most of the younger monkeys soon adopted this behavior. Eventually, the monkeys switched to washing their potatoes in the sea, perhaps because the salt improved the taste. When the researchers began to scatter wheat on the sand, the same young female learned to throw handfuls of wheat mixed with sand into the sea and collect the seeds off the water surface—a behavior also quickly acquired by most of the youngsters. Finally, the researchers began to throw peanuts into the sea, and the monkeys quickly learned to wade into the water to retrieve this favorite food. Astonishingly, many of the monkeys soon took to the water for its own sake, wading, swimming, and diving off rocks even when no food reward was offered. The monkey troop thus began to use an entirely new habitat (figure 5.4).

Figure 5.4. A Japanese macaque washing and eating a sweet potato while her baby looks on. This behavior arose as an individual innovation and then spread "vertically" from mothers to their offspring as well as "horizontally" between unrelated individuals. (© Cyril Ruoso / Biosphoto / Steve Bloom)

An interesting example of an ecological strategy transmitted via social learning has also been described in a nonprimate mammal—the bottlenose dolphins of Shark Bay in Western Australia. Dolphins forage by probing the sea floor with their pointed rostrum, and researchers were intrigued to observe some dolphins wearing conical sea sponges on their rostra as apparent aids in their search for food. Genetic analysis later showed that all the dolphins using this peculiar foraging tool were closely related and, in particular, shared the same mitochondrial genotype. Since mitochondria are inherited almost exclusively via the egg cytoplasm, this means that all "sponging" individuals belong to the same matriline, and strongly suggests that mothers impart this behavior to their offspring by demonstration and learning.[194]

It's not too surprising to discover a capacity for fairly sophisticated social learning in large-brained mammals, but recent studies have revealed the potential for analogous processes in many other animals. In the small Australian skink *Eulamprus quoyii*, males learned the trick of opening a dish containing a mealworm by watching other lizards do it.[195] This shows that the skinks are capable of social learning, but it is not clear how often such transmission might occur within families in a species lacking posthatching maternal care. However, in many species of songbirds, male chicks acquire all or part of their characteristic song through learning and imitation (typically from their father), allowing for the development of local song dialects. Likewise, female chicks undergo "sexual imprinting" on their father's song, learning to prefer males that sound like their father.[196] Social acquisition of sexual preferences has even been reported in fish.[197] A cross-fostering study revealed that stickleback females can sexually imprint on the color and smell of their father. Such behavioral transmission of male sexual signals and female preferences could contribute to reproductive isolation between neighboring populations.[198]

The most compelling demonstration of the spread of a behavioral tradition via social learning in a nonprimate comes from a recent study of the humble great tit (*Parus major*). Great tits and other small birds have one major advantage over lab rats: large populations of individually tagged birds can be studied in the wild over multiple generations, providing precious insights into the behavior and reproduction of wild animals. Taking advantage of a well-studied population of great tits in the Wytham Woods in Oxfordshire, an international team of researchers led by Ben Sheldon at Oxford University devised an ingenious experiment: A few birds from different parts of the woods were brought into the lab and taught to open a box containing a beak-watering delicacy (live mealworms). Importantly, they were taught to open this box in one of two equally effective ways, either moving the blue side of the sliding door to the right, or the red side of the sliding door to the left. Armed with this valuable trick, the birds were then released back into the wild, where the researchers simultaneously set up several snack boxes just like the ones the birds had encountered in the lab. The researchers wanted to know whether other birds would learn to retrieve the mealworms and, most importantly, whether they would imitate the technique used by the savvy lab-trained birds.

They found that, not only did three-quarters of the population quickly learn to open the boxes, but they overwhelmingly acquired the technique of the lab-trained birds that had been released in their part of the woods, so that in one area most birds opened the door from right to left while in another area most birds opened the door from left to right. Moreover, the spread of this tradition through the population followed the established social networks, showing that birds tended to learn from their relatives and friends. Intriguingly, most migrants between areas quickly switched to the locally popular technique, abandoning their own experience in favor of the local fashion and demonstrating social conformity. The researchers removed the boxes after twenty days but replaced them nine months later, by which time 60 percent of the original birds had died. Astonishingly, not only did the old birds remember their preferred technique, but the behavior spread very rapidly among the new generation of birds. Clearly, culture-like traditions can arise and spread via social learning even in some small, short-lived animals.[199]

FELLOW TRAVELERS

The story of heredity is poised for a new twist with the recent recognition that all eukaryotic organisms are actually communities made up of many different species. Our bodies are literally filled with microbes. Bacteria, archaea, and single-celled eukaryotes coat our skin, fill our intestines, and teem in our mouths. We and our microbial partners are also hosts to a vast diversity of viruses. Not to be forgotten are the multicellular symbionts that inhabit the bodies of all large animals—tiny mites and insects that cling to our hair, nematodes, flatworms, and tapeworms that slither and squirm through our intestines and blood vessels. All these distantly related organisms engage in intimate ecological interactions with their host and with each other, and occasionally even share genes through "horizontal gene transfer."[200] The ubiquity and importance of these fellow travelers has led some biologists to suggest that the notion of the solitary individual must be replaced with the concept of the holobiont—a small-scale ecosystem comprised of myriad distinct interacting entities with partially shared interests. Whether we accept the holobiont concept or not, mounting evidence suggests that symbionts are transmitted from parents to their offspring and can influence offspring development in important ways.[201]

The microbiota can influence many aspects of the host's phenotype and ecology. For example, the invasive ladybird beetle *Harmonia axyridis* appears to outcompete native ladybird beetles in part because the invader carries with it a single-celled fungal symbiont that functions as a "biological weapon" against the native species. In many insects, the bacterial endosymbiont *Wolbachia* can induce parthenogenetic reproduction in females, kill sperm, or feminize male embryos, in order to promote its own transmission via eggs. A recent study in *Drosophila* even suggests that bacteria acquired from food influence the fly's sexual signals, to the extent that individuals reared on different media may fail to recognize each other as mates. And, of course, many organisms are almost completely dependent on their microbial partners: termites rely on intestinal protists to digest the cellulose in their woody diet, leaf-cutter ants transport on their bodies the spores of fungi that they cultivate for food, coral polyps host photosynthetic algae, lichens are formed through the intimate association of a fungus and an alga or cyanobacteria, and many plants rely on nitrogen-fixing bacteria within their root nodules, to mention just a few examples.[202]

In our own species, we are accustomed to thinking about the microbes in our bodies as harmful, but many normally have neutral or even beneficial effects. Indeed, the complex chemical activities of this microscopic menagerie are increasingly seen to play key roles in the development, phenotype, and health of their hosts. Recent studies have established strong links between the bacterial ecosystem within our intestines and our metabolism and health. In fact, medical researchers are now even using these principles to treat stubborn illnesses. For example, *Clostridum difficile* is a spore-forming bacteria that is commonly found in soils, but it can also cause opportunistic infections in the gastrointestinal tract of humans. Such infections are routinely treated with antibiotics, but upwards of 35 percent of treated patients relapse once treatment has stopped. These recurring bouts of infection can be life threatening, and they had proven remarkably difficult to cure. That is, until someone had the peculiar idea of transplanting fecal samples from a healthy donor into an infected patient. Remarkably, this cured the patient, and there is now a growing list of cases where doctors have used this approach to great effect.[203] Although the precise mechanism by which it works is not completely clear, the working hypothesis is that a person's normal intestinal microbiota is usually able to outcompete any *C. difficile* bacteria that enter the gut. If an infection does take hold though, then the patient is given antibiotics, and these drugs not only kill off *C. difficile* but the normal intestinal microbiota as well. As a result, once treatment has stopped, if any *C. difficile* remain, they are then free from competitors and so can reinvade the intestine. Only once the competitively superior normal microbiota are reintroduced through a fecal transplant is the intestinal *C. difficile* population driven to extinction, thereby curing the patient.

An even more intimate association exists between complex organisms and the mitochondria (and, in plants and some other eukaryotic lineages, also chloroplasts) that inhabit the cytoplasm of their cells. These organelles are highly modified descendants of bacteria. The ancestors of mitochondria invaded a very distantly related type of single-celled organism (called an archaeon) billions of years ago, ultimately giving rise to the composite eukaryotic cell. Chloroplasts are descended from a different type of bacteria that later invaded eukaryotic cells. In fact, each mitochondrion and chloroplast still has its own circular strand of DNA and replicates by binary fission just like a bacterial cell. Like the bacteria that make up the microbiome, mitochondria and chloroplasts are therefore independently reproducing entities subject to their own genetic and nongenetic inheritance. However, from the perspective of the "host" cell and multicelled organism, the mitochondria and chloroplasts can be viewed as especially well-integrated symbionts, and their transmission across host generations can therefore be regarded as an instance of nongenetic inheritance.

These organelles play essential roles. The mitochondria conduct cellular respiration and metabolism, including the citric acid cycle that produces ATP, the key form of energy storage inside the cell. Chloroplasts use chlorophyll (the pigment that gives plants their green color) to carry out photosynthesis, a chemical reaction that uses solar energy to power the conversion of carbon dioxide into sugar. These organelles are transmitted from mother cell to daughter cells along with the cytoplasm. They are also transmitted from parent to offspring through the germ line, but transmission is usually uniparental (mitochondria are typically transmitted in the egg, but chloroplasts are transmitted either maternally or paternally in various plant taxa). Given their extremely important roles in energy processing within eukaryotic cells, it would be surprising if variation in the features of these organelles did not influence cell function. Indeed, it is now clear that important traits such as metabolic rate, aging, longevity, and even behavior can be influenced by the strain of mitochondria present in an individual's cells.[204]

All of these fellow travelers rely on their ability to cross the generational boundary and infect the offspring of their hosts. Intracellular symbionts and parasites like mitochondria and *Wolbachia* can simply hitch a ride inside the egg, while larger symbionts like mites and worms must generally wait for the offspring to be born. The gut microbes present an especially interesting case. In humans, recent evidence shows that gut bacteria are transmitted from mother to child during normal birth (although there is intriguing evidence suggesting that some of these bacteria manage to invade the womb and infect embryos before birth).[205] Like coprophagy in some mammals and insects, these modes of transmission enable human infants to be inoculated with beneficial maternal symbionts. Mammalian mothers also transmit a variety of skin and nonskin bacteria and viruses to their offspring via milk, perhaps affecting the development of the infant immune, digestive, and even nervous systems.[206] Accumulating evidence suggests that the composition of the intestinal microbiota has a profound effect on an individual's intake and processing of energy and nutrients. An unhealthy diet can alter the bacterial community, and this change in the microbiome could contribute to poor health. The transfer of the microbial ecosystem from mother to offspring may represent an important mechanism whereby a mother's health (including obesity) influences the health of her offspring. Recent evidence even shows that bacteria can be transferred from males to females via semen and physical contact during mating, potentially providing a route for transmission of the paternal microbiome to offspring as well.[207] If the individual is really a multispecies community, then it follows that parent-offspring transmission of the microbiome must be regarded as a distinct dimension of heredity.

THE PARENT AS A TEMPLATE

At the end of chapter 3 we suggested that reproduction can be viewed as a process of autonomous offspring development superimposed upon a parental scaffold. While we believe that the scaffold is a useful general

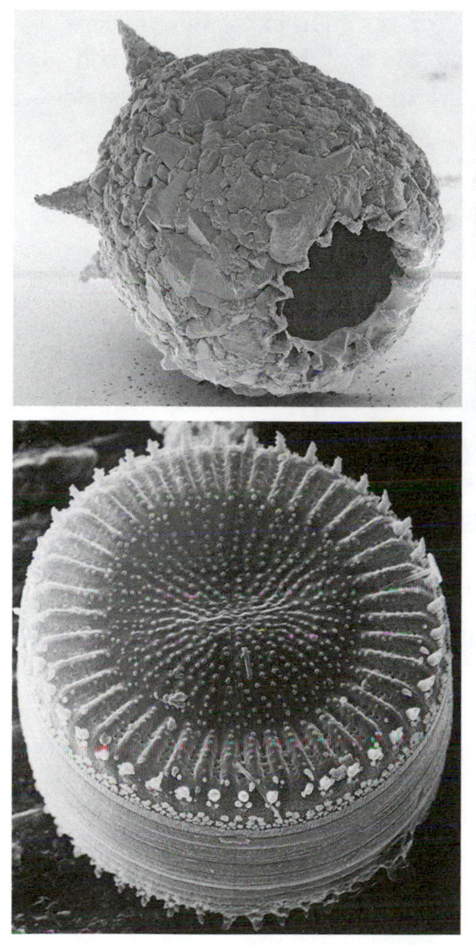

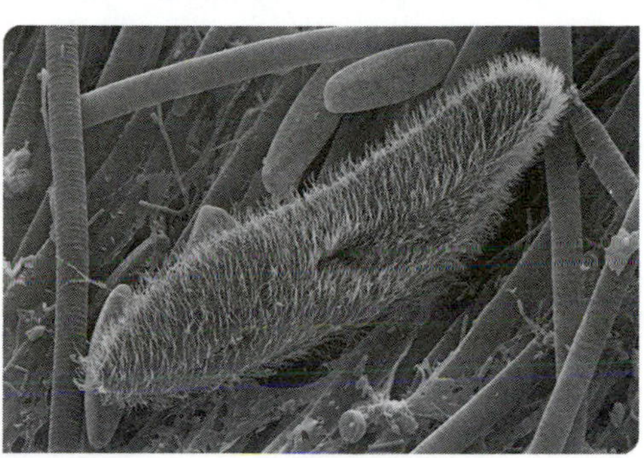

Figure 5.5. The single-celled eukaryotic organisms *Difflugia corona* (*above left*), *Paramecium* sp. (*above*), and *Cyclotella meneghiniana* (*left*) are some of the species in which structural inheritance has been shown to occur. When a cell divides, aspects of its cytoplasmic and membrane structure are transmitted nongenetically to its daughter cells through processes analogous to templating. (*Above left*: © The Trustees of the Natural History Museum, London. *Above*: Wikimedia Commons/Kaden11a. *Left*: © Professor Anne Smith/Science Photo Library)

metaphor for a parent's nongenetic influence on the development of its offspring, some aspects of offspring development are so intricately shaped by parental features that parental influence can be compared not just to a scaffold but to a fine-grained template.

Single-celled eukaryotes such as ciliates and amoebas (collectively called "protists") are tiny but remarkably complex organisms, sporting various combinations of cilia arranged in rows or clumps, propeller-like flagella, specialized "mouths" and "anuses," or even beautiful and intricate calcareous shells (figure 5.5). Several decades ago, protist researchers began to notice that, within clonal cultures made up of genetically identical cells, a great deal of variation exists in the shape and arrangement of features on the cell "cortex," the surface layer consisting of the cell membrane and underlying proteins. Moreover, single cells isolated from such cultures could faithfully perpetuate their cortical variation across many cycles of cell division.[208] A series of exquisite experiments established the role of a nongenetic mechanism—structural inheritance—in the transmission of these features.

The amoeboid protist *Difflugia corona* has a toothed "mouth" that, under high magnification, bears an uncanny resemblance to the terrifying gape of a great white shark. In 1937, Herbert Jennings knocked out some of its teeth using a fine glass needle and discovered that the manipulated cells gave rise to daughter cells with a similarly altered number of teeth, and associated changes in tooth size and shape. By this procedure, he was able to produce new lines of *Difflugia* that differed in tooth number and form.[209]

Three decades later, Janine Beisson and Tracy Sonneborn at Indiana University showed that experimental surgery on the ciliate *Paramecium aurelia* also resulted in stable inheritance of the altered cell forms. *Paramecium* engage in a primitive form of sex called conjugation. When Beisson and Sonneborn caught them in the act and removed part of one of the conjoined cells, the complete cell incorporated the remainder of its partner into itself, resulting in an altered pattern of ciliary rows on the cell surface. The altered cells then passed on their modified cortical structure to their descendants via both sexual and asexual reproduction.[210] Although such experimentally induced variations were sometimes gradually lost, in many cases the modified structure persisted for hundreds of cell generations, generating vast numbers of descendants sporting peculiar ciliary arrangements and altered swimming patterns.[211]

Structural inheritance has since been demonstrated in other single-celled eukaryotes, such as the beautiful calcareous alga *Cyclotella meneghiniana*[212] and the flagellate *Trypanosoma brucei*, the parasite that causes sleeping sickness.[213] This work has not only shown that structural inheritance is extremely widespread across eukaryotic lineages but also highlighted how much is still unknown about the mechanisms involved. For example, research on distant relatives of *Paramecium*, the hypotrich ciliates, has revealed that in addition to the hereditary role of visible structures such as the arrangements of cilia on the surface of the mother cell, some features of the daughter cells' structure are determined by "ultrastructurally unidentifiable" (that is, invisible) features. When stressed, these organisms can undergo "encystment," forming an apparently featureless ball that remains dormant until conditions improve. When surgically modified cells encyst (losing their visible features) and then "excyst" to regain the active cell form, aspects of their modified structure are retained. This points to the existence of a subtle and poorly understood mechanism of structural memory in these cells.[214]

The experimental mutilations that generated striking morphological and behavioral changes in *Difflugia*, *Paramecium*, and other single-celled eukaryotes clearly did not alter any genes but simply modified the structure of the cell, and these acquired traits were often passed on faithfully for many generations. These experiments therefore showed beyond any doubt that, when a cell divides to form two daughter cells, the cortex of the old part of the cell guides the synthesis of the new cortex in something like a templating process. (It's interesting to note the parallels between these experiments and August Weismann's famous mouse tail-severing experiment, discussed in chapter 3. Perhaps the history of heredity would have unfolded very differently if Weismann had set out to investigate the inheritance of mutilations in protists instead of mice.)

But is structural inheritance unique to single-celled eukaryotes, and limited to the arrangement of specialized cortical structures like teeth and cilia? These questions have yet to be answered definitively, but the possibility of structural inheritance in multicellular organisms like ourselves is not far-fetched.

Although evidence of structural inheritance comes mainly from protists, many cellular processes and structures are highly conserved in all eukaryotes and even all cellular life-forms, raising the possibility of structural inheritance in multicelled organisms as well.[215] Indeed, Sonneborn, Beisson, and other researchers have argued that many aspects of cell structure are not (and perhaps cannot be) encoded entirely by DNA sequences and must instead be transmitted via template-like processes.[216] Historian Jan Sapp summarized this principle very clearly: "Cells never arise by aggregation, but by the growth of pre-existing cells; they model themselves upon themselves. As a cell grows and makes itself, the macromolecules specified by the genes are released into a context that is already spatially structured."[217] In other words, the components of a cell are not capable of self-assembly; preexisting structures like cell membranes and cytoskeletons are required to give form to new cells and to express the genetic instructions contained in the DNA sequence. As Thomas Cavalier-Smith put it, "there is an important difference between a chicken and chicken soup." The role of structural inheritance in heredity should therefore depend on the extent to which variations in preexisting structure can be transmitted from cell to cell or, in the context of inheritance across generations of multicellular organisms, the extent to which structural information contained in the gametes can affect the development of embryonic features and can be transmitted to the embryonic germ cells.

As we saw in chapter 3, it has been recognized for a long time that some aspects of early development depend on the structure of the egg cell. For example, early experiments in which the large eggs of animals such as sea urchins and sea squirts were centrifuged to disturb the cytoplasm's structure resulted in a host of developmental abnormalities, such as embryos that seemed to be turned inside out. Embryologists eventually established that some aspects of development reflect the structure of the egg cell's cortex[218]—a complex outer layer that shares many features with the cortex of single-celled eukaryotes like *Paramecium*. Likewise, the egg's polarity, the nature, quantity, and location in the egg cytoplasm of biomolecules such as proteins and RNAs, and perhaps even the distribution within the cell of various organelles, all contribute to the regulation of patterns of cell cleavage in the early embryo, the establishment of distinct cell lineages, and the epigenetic reprogramming and activation of the embryonic genome.[219] In other words, key aspects of early development depend on the structure of the egg, which is built by the maternal body and whose features are therefore subject to a range of maternal effects that could reflect influences of both the maternal genome and the maternal environment. Although many aspects of egg structure are probably determined by maternal genes, even such genetically determined features are transmitted nongenetically from the mother to her offspring (that is, as indirect genetic effects), and understanding their evolutionary consequences therefore requires the tools of extended heredity. Moreover, research on mice by Karolina Piotrowska and Magdalena Zernicka-Goetz at Cambridge University has shown that some of the earliest developmental stages after fertilization are patterned according to the precise location where the sperm penetrates the egg,[220] raising the possibility that variation in features of the sperm cell could also influence the course of development.

The intracellular organization of eggs and other eukaryotic cells appears to depend largely on the structure of the cytoskeleton—a complex, asymmetrical framework of microtubules that link the inner cytoplasm to the outer cortex.[221] In animals and many other eukaryotes, the microtubules radiate from an organelle called the centrosome, which is composed of two gear-like structures called centrioles arranged at right angles and surrounded by a layer of proteins. Interestingly, in a fertilized animal egg, the centrosome is assembled from both maternal and paternal components—a complete centriole from the sperm and various proteins from the egg. The architecture of the cytoskeleton may, in turn, influence the locations of various organelles within the cytoplasm, the way a house's architectural layout influences the location of its contents. Importantly, because the cytoskeleton is intimately linked to the cortex, modifications to the cortex like those Beisson and Sonneborn induced are likely to alter the structure of the cytoskeleton as well, and the hereditary transmission of such modifications from mother to daughter cells may therefore reflect the transmission of cytoskeletal form.[222]

The structure of the molecular envelope that encloses the cell cytoplasm and cytoskeleton—that is, the cell membrane—also appears to be transmitted nongenetically from mother to daughter cells. Thomas Cavalier-Smith has pointed out that cell membranes can form only as extensions of existing membranes, so that all existing cell membranes can be traced back to the first cell. And, because new membranes are assembled on the pattern of the maternal membrane, cell membranes can embody hereditary information in their own right (although membrane heredity appears to be limited to just a few distinct membrane types).[223]

Beyond these clues, we still know little about the potential for maternal and paternal nongenetic influence on the structure of zygotes, the extent to which such influences might translate into variation in development, or the potential for nongenetic transmission of structural variation beyond a single generation in multicelled organisms. If stochastic or environmentally induced structural variations in the zygote are to influence adult features, these structural features must be transmitted to somatic cell lines within the body. If such effects are to persist beyond a single generation, zygote structure must influence the structure of the germ line cells that arise through multiple cycles of cell division. Such effects are possible in principle, but have not yet been demonstrated. Such effects remain poorly known because of the considerable challenges inherent in characterizing variation in the three-dimensional structure of cells. The discovery of structural inheritance in *Paramecium* and other single-celled eukaryotes was facilitated by the relatively large size of these cells, their ease of culturing, and the ability to visualize their complex cortical topography using silver staining techniques. The development of more powerful techniques for research on cell structure in multicelled organisms may reveal a hitherto-unappreciated role for structural inheritance in complex eukaryotes as well.

PROTEIN MEMORIES

There is another form of structural inheritance that could play a role in both single-celled and multicelled eukaryotes—the capacity of certain proteins, called prions, to impose their three-dimensional shape on other proteins. Prions were originally discovered by researchers searching for causes of a mysterious neurodegenerative disorder called kuru that devastated certain communities in remote areas of Papua New Guinea. Kuru bedeviled scientists because it was transmitted within families as well as to female (but not male) in-laws, a very odd pattern of inheritance that defied both genetic and environmental explanations. The illness was eventually traced to the cultural practice of eating the brain of deceased loved ones, a ritual in which biological relatives as well as female (but not male) in-laws usually participated. The role of an infective agent was confirmed when researchers showed that tissue from human victims could infect chimpanzees. As in the case of cortex structure inheritance in protists, researchers spent years searching for a DNA-based or viral mechanism of transmission. Instead, it was eventually shown that the hereditary agent was a protein that could somehow impose its three-dimensional folding structure on other proteins with similar amino-acid sequences. A number of other neurodegenerative prion diseases have since been described in humans and other mammals, including Creutzfeldt-Jakob disease, which runs in families but can also be acquired from contact with brain tissue, and "mad cow" disease, which can be transmitted from cows to humans through the consumption of beef.

Despite the notoriety of prions as agents of disease, prions and other proteins with prion-like properties occur in both single-celled and multicelled eukaryotes and appear to play essential physiological roles.[224] For example, prions in the mouse brain may function in protection of neurons from age-related degeneration, as well as in the formation and preservation of long-term memories.[225] Prions could also contribute to heredity. In yeast, prions have well-known roles as cytoplasmic hereditary factors that can respond to environment and modify patterns of gene expression.[226] Interestingly, some prions have the capacity to take on and transmit several different three-dimensional structures. Because a protein's biological properties are determined in large part by its three-dimensional folding structure, the different structures assumed by prions could have different physiological effects. Many more prion-like cytoplasmic factors may await discovery.[227]

THE CHALLENGES OF UNCOVERING NONGENETIC INHERITANCE

Research on nongenetic inheritance presents a number of difficulties. As we have seen, nongenetic inheritance is often detected as an effect of parental environment on descendants. In principle, the experiments required to detect such effects are quite simple.[228] For example, if one wants to know whether elevated fat intake can affect the health and longevity of offspring and grand-offspring, one can feed groups of experimental animals a high-fat diet while feeding control animals a normal diet, breed them with partners fed on a normal diet, and then observe their offspring. But all such studies face a complication: a potential for the confounding effects of genes. Imagine that some animals are more susceptible to the harmful effects of a high-fat diet because of the genetic alleles they carry. Such animals will be more likely to die before reproducing, or to produce fewer surviving offspring, and this could mean that natural selection will bias the genotypes of offspring from the fatty and control diet treatments in the experiment. Any differences in health or longevity between these offspring could then result from genetic differences rather than from nongenetic factors passed down by their parents. The potential for such confounding effects can be minimized by using isogenic or highly inbred lines, or by splitting families and subjecting siblings to contrasting environments. Individuals subjected to different experimental treatments will thus have similar genotypes, providing little genetic variation for natural selection to act on. Even so, the possibility of confounding genetic effects can still crop up because stress can elevate the genetic mutation rate.[229] In other words, even if genetically identical animals are used, the potential for a high-fat diet to induce new mutations in the germ-line can mean that offspring of parents fed a fatty diet will have more mutations than offspring of parents fed a normal diet. Fortunately, genetic and nongenetic effects can often be distinguished based on their characteristics. For example, nongenetic effects are likely to be more consistent across individuals; most animals fed a high-fat diet might produce offspring with a particular metabolic syndrome, while genetic mutations caused by high-fat diet are unlikely to be as common or as consistent in their effects on offspring. Nongenetic effects are also likely to disappear relatively quickly once the environmental trigger is removed, whereas genetic mutations might be transmitted over many generations, until eliminated by natural selection.

But there is another complication. The great majority of studies on nongenetic inheritance have taken the approach of contrasting just two environmental states, such as "normal" versus "fatty" diet or "normal" versus "hot" rearing temperature. Such studies can produce misleading results when the effects of parental environment are nonlinear, because in such a case the direction of the effect, and even whether an effect is detected at all, might depend on the specific environments used—a choice that can be rather arbitrary (after all, what is a "normal" diet for a fly or a mouse?). Here, investigators of nongenetic inheritance can borrow from research on developmental plasticity, where it has become standard practice to investigate more than two environments, and sometimes more than one environmental factor, within the same study. Nutritional scientists have developed a particularly useful approach known as "nutritional geometry."[230] Under the geometric framework, the amounts of two or more dietary nutrients are manipulated simultaneously, with the lowest and highest nutrient concentrations spanning the extremes that animals will tolerate. For example, researchers interested in the effects of eating protein and carbohydrates will create diets that vary in both the ratios and total concentrations of these nutrients, resulting in twenty or more different diets. They will then raise animals on each of those diets and assess them for traits such as health, longevity, or total fitness. This approach makes it possible to detect nonlinear effects of ingesting each nutrient, as well as complex interactions between the effects of the different nutrients. The geometric approach readily lends itself to research on the effects of ancestors' diet on descendants and can be extended to investigate the effects of other factors such as social environment or temperature.[231]

Once a case of nongenetic inheritance has been uncovered, the next challenge is to identify the proximate mechanism involved. Of the examples of nongenetic inheritance that have been reported, we can point to few cases for which we have a reasonably complete understanding of the chain of cause and effect mediating the influence of the parent on its offspring. In most cases, this physiological and biochemical cascade is likely to be very complex, involving many steps in the parent as well as the offspring.[232] Yet, it's the steps in this chain that largely determine the properties of the inheritance mechanism, such as its transgenerational stability and capacity to transmit effects of ancestral environment.

One challenge is to understand the link between the body's reproductive and nonreproductive tissues. For example, how does a psychological response to stress ultimately manifest as an epigenetic change in eggs or sperm? Although hormones and regulatory RNAs are well established as factors mediating soma to gamete transmission in plants, the nature of the link between soma and germ line remains poorly understood in animals.[233] Soma-to-germ line transmission must also be distinguished from cases where an environmental factor affects both somatic and reproductive tissues simultaneously. For example, a toxin might spread through the bloodstream and infuse all tissues including the germ line.

A related challenge is to uncover the factor actually transmitted across generations—that is, the causal link between parent and offspring. Given the vast variety of different factors passed to offspring in the gametes and, in some species, the equally vast diversity of postfertilization influences of parents on the development of their offspring, attempts to identify this key hereditary factor can be likened to the proverbial search for a needle in a haystack. As we have seen, similar inheritance patterns can arise from different mechanisms. For example, a maternal effect in a mammal could be mediated by epigenetic or structural changes in eggs, by altered conditions in the uterus, or by changes in postpartum factors such as maternal behavior or milk composition.[234]

Rather than starting with a heritable phenotype and seeking to identify the inheritance mechanism involved, some researchers have attempted to take the opposite route—scanning for variation in potential hereditary factors and then attempting to infer their possible phenotypic role. For example, a number of recent studies have sampled epigenetic variation in natural populations. In the genetic realm, this is equivalent to a search for variable DNA sequences, yielding a list of single-nucleotide polymorphisms (SNPs) whose functional significance (if any) would have to be determined through follow-up studies. Challenging as such an approach would be in relation to genetic variation, when it comes to epigenetic variation the difficulties are even greater.[235] To begin with, one must establish that the epigenetic variation in question is independent of genetic variation. Some of the epigenetic variation detected by epigenome sequencing will simply result from DNA-sequence variation, since a genetic allele may induce an epigenetic change in itself or in another part of the genome as part of the biochemical cascade linking genotype to phenotype.[236] Evidence that epigenetic variants are partly or wholly independent of genes can be obtained via experimental manipulation of environment,[237] by associating epigenetic changes across cohorts within a population with fluctuation in ambient conditions such as food availability,[238] or by manipulating the epigenome directly (see chapter 8). An epigenetic variant induced by environment is either of the "facilitated" or "pure" variety, and therefore potentially transmissible to offspring. However, while we can assume that a genetic allele or SNP can be transmitted across generations, we cannot be sure that a given epigenetic variant can be transmitted through the germ line because much of the epigenome is reprogrammed between generations. Only a subset of epigenetic variation can make it across generations in multicellular organisms, and very little is known about which particular epigenetic variants can do so. Third, just as a genetic SNP need not be causally related to any phenotype, an epiallele need not necessarily have any functional consequences.

A further complication is that the environment can sometimes have direct and repeatable effects on the DNA sequence itself. The CRISPR-Cas system (which we will discuss in more detail in chapter 10) is an example of such modifications in bacteria, and recent evidence suggests that environment can induce repeatable DNA sequence changes in multicellular organisms as well. In *Drosophila*, diet can alter the number of copies of genes coding for a component of the ribosome—the factory-like organelle where messenger RNA is translated into protein. In animals, ribosomal genes are present in multiple copies arranged in tandem repeats. Flies fed a diet high in yeast were found to not only lose copies of ribosomal genes but also to transmit their altered genomes to their descendants for multiple generations.[239] Likewise, telomeres—repetitive sequences found at the tips of chromosomes—can shorten in response to environmental stress, and offspring can inherit the telomere lengths of their parents.[240] These phenomena involve transmission of a nuclear DNA sequence or the number of repeats in a repetitive sequence and therefore do not quite fit within our definition of nongenetic inheritance (that is, inheritance via transmission of factors other than DNA sequences). However, they can mediate transmission of environmental effects to offspring and could therefore be mistaken for nongenetic inheritance.

WHY DOES NONGENETIC INHERITANCE EXIST?

It should now be clear that a diverse array of mechanisms exists for the transmission of nongenetic information across generations. As we have seen, although some of these nongenetic factors appear to be of adaptive value to the organisms that harbor them, many of them are deleterious. One might therefore be led to wonder why natural selection hasn't disposed of these maladaptive instances of nongenetic inheritance. In this final section, we consider why nongenetic inheritance exists despite these apparent costs.

To begin with, the logic of heredity necessitates the existence of some form of information transmission across generations in addition to genes. This is true simply because the string of digital information embodied by a genome has no meaning except in the context of an appropriate machinery for reading it. If an alien civilization were to send us a string of symbols we would have no way of knowing whether they coded for one of the species found on their planet, for an extraterrestrial bank account, or for nothing at all. One might argue that the machinery for reading DNA could itself also be encoded by the genome, and it's certainly true that many of the cellular constituents involved in organismal development and regulation, such as proteins and RNAs, are synthesized on the basis of genetic instructions. However, this presents us with a classic chicken-and-egg problem: to decode the genetic instructions and synthesize the required cellular constituents, a machinery must already be in place. So there is no escaping the fact that some form of interpretive information must be transmitted alongside the genome, and that this interpretive information must be embodied in a preformed physical machinery. In all cellular life-forms, the machinery takes the form of a membrane-enclosed cell containing a highly structured cytoplasm, and this machinery has been transmitted from cell to cell alongside genes ever since the dawn of cellular life. In many complex organisms, this machinery extends far beyond the single cell to encompass features such as a uterus, milk, parental behavior, or other parental traits that are required for normal offspring development to take place.

Given that an interpretive machinery must be transmitted alongside the genome, it should not come as a surprise that differences exist between individuals in the components that make up this machinery and in the information that they embody. Moreover, as we saw in this chapter and the previous one, there is a great deal of evidence that some components of this machinery vary independently of DNA sequences—that is, individuals differ in their cellular machinery, and in other aspects of their phenotype such as behavior, because of spontaneous or environmentally induced changes in nongenetic factors. A great variety of such nongenetic factors are transmitted to offspring, and some have the capacity to regenerate themselves and persist over multiple generations. The features of the interpretive machinery that are transmitted to offspring are not fully determined by parental genes because living organisms are complex, fragile, and highly sensitive to their environments. Genes have a great deal of influence, but their control is far from absolute.

Likewise, it should come as no surprise that much of the variation in nongenetic components of heredity is deleterious. After all, no one is surprised by the fact that the vast majority of known genetic variants are deleterious, and few people would stop to ask why natural selection has failed to completely eliminate the problem of genetic mutation. Most genetic and nongenetic changes are maladaptive for the same reason: such changes occur as spontaneous errors or accidents that alter—and usually damage—the structure of cellular machinery. This is clearly what happens, for example, when male rat embryos are exposed to endocrine disruptors such as vinclozolin in the womb; these chemicals disrupt epigenetic factors that are then transmitted to descendants.

Of course, just as fitness-enhancing genetic mutations occasionally occur, so too can fitness-enhancing nongenetic changes. For example, as we saw in chapter 4, a diet-dependent epigenetic mark transmitted by rats to their offspring confers the healthy "pseudoagouti" phenotype, and this epigenetic trait can respond to selection. But, as we have also seen, nongenetic factors can play a role beyond simply being a highly mutable and environmentally sensitive substrate for random variation. Many mechanisms of nongenetic inheritance exist, and so we might expect natural selection to act on this variation in inheritance mechanisms as well, preserving and honing those that provide a fitness benefit. For instance, we have seen that certain nongenetic inheritance mechanisms can allow parents to adjust the features of their offspring to suit the anticipated environmental conditions. Thus, natural selection maintains the existence of some nongenetic inheritance mechanisms and co-opts them for adaptive functions, while other instances of nongenetic inheritance occur simply as unavoidable by-products of the reproductive process.

6

Evolution with Extended Heredity

> Selection has been studied mainly in genetics, but of course there is much more to selection than just genetical selection.
>
> —George R. Price, "The Nature of Selection," 1995[241]

In the previous two chapters, we highlighted the overwhelming evidence for the existence of other forms of inheritance in addition to genetic inheritance. To us what is perhaps even more impressive than the sheer volume of this evidence is the scope of variation in the kinds of inheritance mechanisms that exist. As we explained in chapter 2, however, we believe that all these forms of nongenetic inheritance have features in common that distinguish them from genetic inheritance. As a result, we have chosen to delineate just two channels of information transmission across generations: genetic inheritance and nongenetic inheritance. In this chapter and the next we show how these two modes of inheritance can be incorporated into existing evolutionary theory, allowing us to investigate the processes and outcomes of evolution when inheritance involves more than genes. But first we will introduce an influential figure in the history of evolutionary theory whose ideas form the basis of our approach.

THE PRICE OF EVOLUTIONARY THEORY

The history of evolutionary biology is populated with quirky characters and interpersonal feuds that have become the stuff of legend within the

Figure 6.1. George Price developed a way to model evolution that's sufficiently flexible to accommodate both genetic and nongenetic inheritance. Price's equation can be extended to encompass extended heredity's diverse mechanisms of inheritance.

discipline. But the story of George Price must surely stand above them all in its unique and bitter blend of comedy, brilliance, and tragedy.

Price was a US scientist, trained in chemistry at the University of Chicago (figure 6.1). Intellectually restless, he devoted his early career energies to various projects including working as a consultant for the Argonne National Lab on the Manhattan Project, teaching chemistry at Harvard University, and working at the storied Bell Laboratories during the era when luminaries like Claude Shannon, the father of information theory, were also making their scientific mark on the world.[242]

Over time Price developed a very eclectic set of interests, ranging from the geopolitical to the supernatural. One of his first scientific contributions was a refutation of extrasensory perception and parapsychology.[243] However, as Oren Harman relates in his biography of Price, *The Price of Altruism*,[244] in 1966, at the age of forty-four, Price was stricken with thyroid cancer. In a strange turn of events, a surgical error during an operation to remove the tumor left him with lifelong mobility problems. While no doubt devastating, the insurance settlement arising from this failed operation afforded Price the financial independence to move to London, where he embarked on a new set of interests involving evolutionary theory. It was during this period that Price published some of the most foundational papers in the field.

Perhaps Price's most famous work, published in 1970 in the journal *Nature*, was a short paper with the title "Selection and Covariance."[245]

In it he put forth a new mathematical conceptualization of evolution by natural selection. Up until that point several mathematical treatments of evolution by natural selection had been developed, primarily within the field of population genetics. The significance of Price's landmark paper was not that it provided a new theory of natural selection, but rather that it showed how all previous mathematical treatments could be viewed as special cases of a more general conceptualization of selection.

But Price's goal was much more ambitious than this. His intent was not just to subsume previous models from population genetics. Rather, he felt that biological evolution through natural selection was simply one instance of a more general process of selection. The abstraction of selection more generally would presumably then open the door to a greater understanding in many different fields of study. In Price's own words, a "model that unifies all types of selection (chemical, sociological, genetical, and every other kind of selection) may open the way to develop a general 'Mathematical Theory of Selection' analogous to communication theory."[246] Undoubtedly this ambition harkened back to his experiences at Bell Labs in New Jersey.

From this quote, it is clear that Price's motivation was not the same as that which concerns us here. In particular, he was not motivated by the belief that there is more to biological evolution than genetic change. Instead, he felt that there was more to selection and evolution than that which biologists study. But as we will see, because his abstraction of selection was so far-reaching, it also readily lends itself to the goal of formalizing the process of evolutionary change when heredity is mediated by any transmitted entity, be it genetic or otherwise—in other words, evolution with extended heredity.

CALCULATING EVOLUTION

At its core, Price's analysis of selection focuses on an abstract population of entities. These entities might be biological entities like genes, cells, or individual organisms, but they can equally well be manufacturing companies or dialects of a language. Entities can be viewed as reproducing themselves from one point in time to the next, and thus we can talk about an "ancestral" and a "descendant" generation. The properties of the entities can change upon reproduction such that descendant entities need not resemble the parents that produced them. The goal is then to describe, mathematically, how the population of these entities changes from one generation to the next.

To see how this might work, let's consider a fanciful example involving a population of people who enjoy consuming tea. Tea culture has undergone something of a revival in various parts of North America, but not long ago the extent of many North Americans' experience with tea involved scrounging a stale tea bag from the back of the kitchen cupboard when a guest unexpectedly preferred a cup of tea to coffee. Yet the preparation of a good cup of tea can be a much more nuanced affair. Indeed, the ritual of tea preparation is a complex cultural tradition that has been passed down between generations for hundreds of years.

Like most recipes and traditions that are taught to younger generations by elders, tea preferences have slowly changed and diversified over the course human history, displaying a characteristic evolutionary pattern of descent with modification. One relatively recent innovation is the addition of sugar to tea as a sweetener. While this was not customary in China where tea originated, variations of this practice are now quite common in many countries.

To explore the cultural evolution of tea preferences in more depth, let's imagine a population in which some people prefer plain tea while others enjoy sweetening their tea with sugar. In fact, to make things concrete, let's suppose that 4/5 of the population take their tea plain while 1/5 prefer it with sugar. Figure 6.2 displays a schematic of the population.

Now, in each generation the elders of the population teach young people the cultural practice of making tea. In so doing they convey their own preferred practice of either drinking tea plain or adding sugar to sweeten its flavor. Over time, through several generations of cultural transmission, we might expect the typical tea preference in the population to change. For example, a preference for sweet tea might become more common if sweet-tea drinkers tend to teach more young people than plain-tea drinkers (as might have happened at some point after the introduction of tea to Europe in the seventeenth century).

To take this example further, let's simplify matters by supposing that generations are discrete in time and that the number of people who drink tea remains constant across generations. This means that, *on average*,

Figure 6.2. A schematic representation of a population in which four-fifths of the people prefer plain tea while one-fifth prefer sweetened tea.

each tea-drinking elder will teach a single young person their preference for tea. But although this is true on average, some tea-drinking elders might nevertheless teach a greater number of youth than the average while others might teach a smaller than average number.

Let's imagine that those older people with a preference for sweet tea in fact teach 3 times more than the average, while those with a preference for plain tea teach ½ of the average. Since the average is 1, this means that each sweet-tea-drinking elder teaches 3 young people and each plain-tea-drinking elder teaches only 1/2 of a young person (put another way, it takes 2 plain-tea-drinking elders to teach a single young person). As a check, the average number of young people taught by a single elder is given by the faction of sweet-tea-drinking elders (1/5) multiplied by the number of young people that each teaches (3), plus the fraction of plain-tea-drinking elders (4/5) multiplied by the number of young people each of these teaches (1/2). This gives $1/5 \cdot 3 + 4/5 \cdot 1/2 = 1$, as it must if the number of people who drink tea is the same in both generations.

Under these conditions, 3/5 of the young people will be taught by sweet-tea-drinking elders even though these elders make up only 1/5 of their own generation. This is because each of these elders teaches 3 people. On the other hand, only 2/5 of the young people will be taught by plain-tea-drinking elders despite the fact that these elders make up 4/5 of their own generation. This is because each of these elders teaches only 1/2 a person. Figure 6.3 illustrates this idea schematically. Now if each type of elder passes on their tea preference to individuals of the younger generation with perfect fidelity, then 3/5 of the younger population will have a preference for sweet tea while 2/5 will prefer it plain (figure 6.4).

From figure 6.4 we can see that the preference for sweet tea has increased from the ancestral to the descendant generation by an amount of $3/5 - 1/5 = 2/5$. This increase arises from the fact that sweet-tea-drinking ancestors teach a greater fraction of the population than

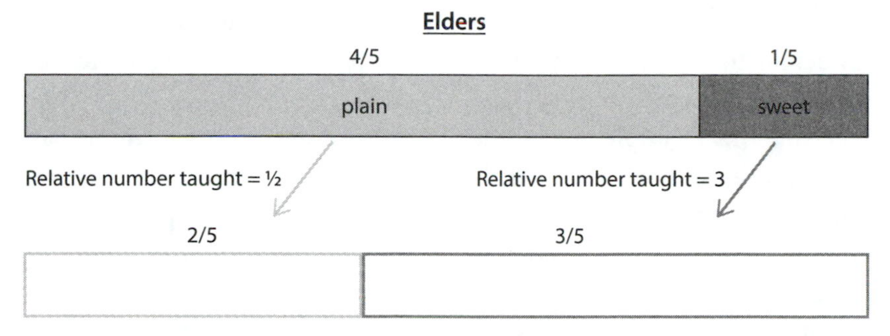

Figure 6.3. A schematic representation of how the preference for different types of tea in a population changes as a result of cultural inheritance. The shading of the outline of each box in the second row indicates the type of elder from which preferences were acquired. Two-fifths of the younger generation acquire their preference from plain-tea-drinking elders while three-fifths acquire their preference from sweet-tea-drinking elders.

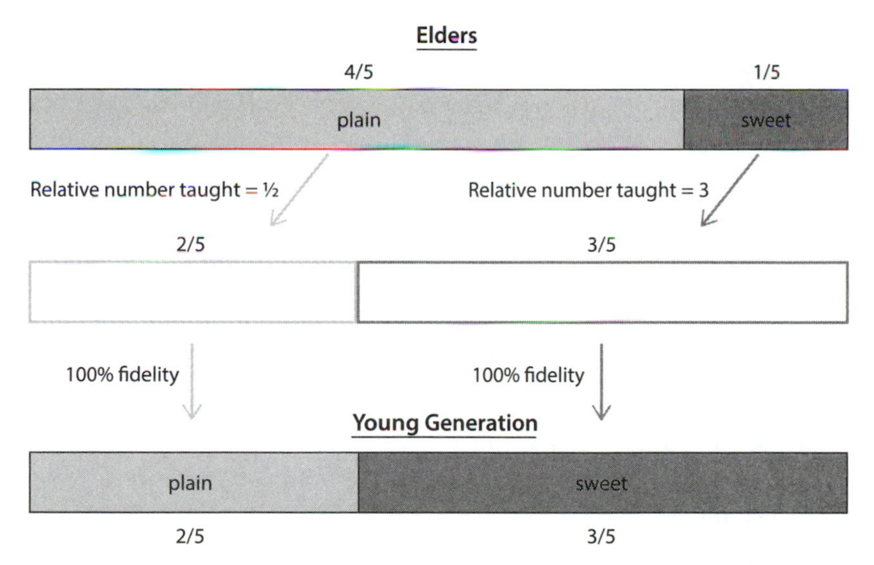

Figure 6.4. A continuation of figure 6.3 that shows how fidelity of learning can be incorporated into cultural evolution. The final row shows the composition of the younger generation if each type of elder passes on their preference with perfect fidelity. Three-fifths of the younger generation will prefer sweet tea.

plain-tea-drinking ancestors. We have thus obtained a very simple model for the evolutionary change of a population via nongenetic, cultural, inheritance.

While this example goes some way toward describing a simple case of cultural evolution, it nevertheless is missing some important features. For example, adolescent rebellion being what it is, some people of the younger generation will undoubtedly want to assert their individuality by rejecting the preference of their elders. Thus, despite an elder's best intentions, the fidelity of transmission of their preference to the younger generation is not going to be perfect. What's more, different preferences might have different fidelities because some preferences might engender a greater degree of loyalty. Perhaps those who enjoy plain tea are stubbornly steadfast in convincing younger individuals to accept their preference while those who enjoy sweet tea are more malleable in their opinion. Or perhaps, returning to seventeenth-century Europe, youngsters taught to make sweet tea are less likely to maintain this practice simply because of the exorbitant price of sugar.

To incorporate such an effect, let's imagine that the preference for plain tea is passed to the next generation with perfect fidelity but that the transmission of the preference for sweet tea is faithful only 2/3 of the time. This means that 2/3 of the time, when a sweet-tea-drinking elder teaches a young person, that young person acquires a preference for sweet tea. The remainder of the time (that is, 1/3 of the time) even though the young person is taught by a sweet-tea-drinking elder, they nevertheless acquire a preference for plain tea. Thus all of the descendants of plain-tea-drinking elders will prefer plain tea, and 1/3 of the descendants of sweet-tea-drinking elders will prefer plain tea as well. Figure 6.5 illustrates this idea.

In this case we see that now 2/5 of the younger generation prefer sweet tea and 3/5 have a preference for plain tea. Just as in figure 6.4, the frequency of the preference for sweet tea has increased in figure 6.5, but it has increased by a smaller amount (a change of 2/5 – 1/5 = 1/5 in figure 6.5 versus a change of 3/5 – 1/5 = 2/5 in figure 6.4). The smaller increase in figure 6.5 stems from the fact that, although each sweet-tea-drinking elder teaches a greater number of youth than each plain-tea-drinking elder, the fidelity of transmission of the preference for sweet tea across generations is lower than that of the preference for plain tea.

PRICE'S EQUATION

The two processes embodied in the simple example of figure 6.5, namely, differential reproduction of types (in terms of the number of people taught) and differential fidelity of transmission of types, are exactly the two processes that Price combined in his elegant mathematical formulation of selection and evolution. The only difference between Price's formulation and our analysis is in its level of generality. Price developed an extremely general description of selection and evolution that can be applied to any problem. This includes examples of the sort of cultural evolution of tea preference described here as well as evolution via any other form of nongenetic or genetic inheritance.

For those readers with a penchant for the mathematical, the derivation of Price's beautiful equation in its full generality is given in Box 6.1.

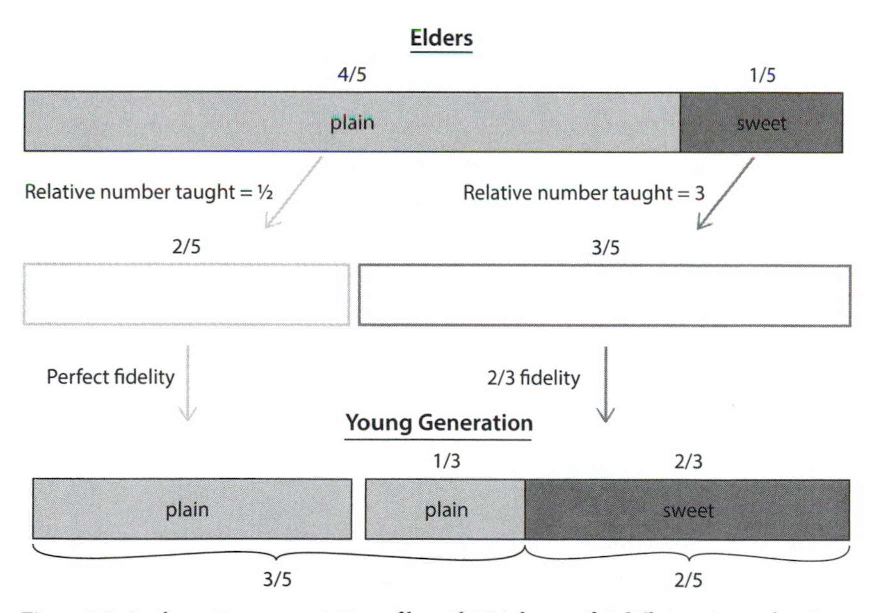

Figure 6.5. A schematic representation of how the preference for different types of tea in a population changes as a result of cultural inheritance that includes differences in teaching as well as differences in fidelity of learning. This figure is analogous to figure 6.4 except that there are now differences in the fidelity of learning. The final row shows the composition of the younger generation if each sweet-tea-drinking elder passes on his/her preference with two-thirds fidelity. Two-fifths of the younger generation will prefer sweet tea.

The equation focuses on the value of an arbitrary trait of interest, represented by the letter z. It then states that the change in the average value of this trait from the ancestral to the descendant generation is given by

$$\text{Change in average } z = \text{cov}[z, w] + E[wd]$$

where w represents an individual's reproductive success, measured as the relative number of descendants produced in the next generation, and d represents the average difference between the trait value of a descendant and that of its parent.

Box 6.1. The Price Equation

The Price equation can be derived directly using elementary algebra. Suppose a population of individuals is subdivided into types indexed by i, with z_i being the value of some trait of interest for individuals of type i (for example, their tea preference). Now use q_i to denote the fraction of the ancestral population made up of individuals of type i. The average value of z (denoted by $\bar{z}$) in the ancestral population is therefore $\bar{z} = \sum_i q_i z_i$.

In a similar way let's use q'_i to denote the fraction of the descendant population that "descends from" ancestors of type i. The average value of z in the descendant population is therefore $\bar{z}' = \sum_i q'_i z'_i$ where z'_i is the average trait value of descendants of ancestral individuals of type i. Putting these two results together, the change in the average value of z from one generation to the next is therefore

$$\Delta\bar{z} = \sum_i q'_i z'_i - \sum_i q_i z_i$$

To simplify this equation, we can first make use of the fact that $q'_i = \frac{q_i W_i}{\bar{W}}$, where $\bar{W} = \sum_i q_i W_i$ is the average reproductive output in the parental generation. To see this more clearly, suppose there are N_{Total} individuals in the population. The number of individuals of type i in the ancestral population is therefore $N_i = q_i N_{Total}$. Each of these individuals produces a total of W_i descendants, and therefore, the fraction of the descendant population that comes from type i ancestors is

$$q'_i = \frac{N_i W_i}{\sum_i N_i W_i} = \frac{\frac{N_i}{N_{Total}} W_i}{\sum_i \frac{N_i}{N_{Total}} W_i} = \frac{q_i W_i}{\sum_i q_i W_i} = \frac{q_i W_i}{\bar{W}}$$

In fact, to simplify the notation further, let's write $w_i = W_i/\bar{W}$. In words, W_i is the reproductive output of an individual of type i and w_i is its reproductive output *relative to the average reproductive output in the population as a whole*. Second, we notice that the value of z in descendants of parents of type i can be written as $z_i' = z_i + d_i$. Putting this together we get,

$$\begin{aligned}\Delta\bar{z} &= \textstyle\sum_i q_i' z_i' - \sum_i q_i z_i \\ &= \textstyle\sum_i q_i w_i (z_i + d_i) - \sum_i q_i z_i \\ &= \textstyle\sum_i q_i w_i z_i - \sum_i q_i z_i + \sum_i q_i w_i d_i\end{aligned}$$

The first two terms can be written as the covariance $\text{cov}(z, w)$ and the second term is the average or expected value of wd, which can be written as $E[wd]$. Thus we have

$$\Delta\bar{z} = \text{cov}(z, w) + E[wd] \qquad (1)$$

Before we begin to unpack the somewhat daunting-looking notation in Price's equation, let's first consider how it relates to our example of tea preferences. After all, in our example we focused on how the *frequency* of the preference for sweet tea changed from one generation to the next and not on how the *average value* of some trait changed during this time. But with a bit of insight we can make a direct connection between the two. First, let's define a binary trait whose value z represents an individual's preference for sweet tea. Specifically, let's suppose that an individual's value of the trait is $z = 1$ if they prefer sweet tea or $z = 0$ if they prefer plain tea. If q_s is the frequency of the preference for sweet tea in the population, then the average value of the preference trait in the population is calculated to be

$$\text{average } z = q_s \times 1 + (1 - q_s) \times 0 = q_s$$

In words, the average value of the preference trait is nothing other than the frequency of the preference for sweet tea! Thus, our example of the cultural evolution of the frequency of tea preferences fits squarely with the framework of Price's equation as a special case.

With this insight in hand, let's now decipher Price's notation. The quantity $\text{cov}(z, w)$ in his equation represents the *covariance* between the value z of the trait and the relative reproductive output w across individuals in the ancestral population.[247] If individuals with a large value of z also tend to have a large reproductive output (that is, a large value of w), then this covariance will be positive. All else equal this will thereby tend to make the change in the average value of z positive (that is, the average value of z will increase).

Returning to our tea example, individuals with a preference for sweet tea (that is, those who have the "high" value of $z = 1$ rather than $z = 0$) tend also to have a larger reproductive output, meaning that they teach more youngsters. Thus, large values of z are associated with large values of w, making the covariance term in Price's equation positive. All else equal this therefore tends to increase the average value of z across generations (that is, the frequency of the preference for sweet tea increases). Thus, the covariance term of Price's equation captures the effect of differences in the *amount* of reproduction among types.

The quantity $E[wd]$ in Price's equation represents the expected, or average, change between parent and descendant in the value of z as a result of imperfect fidelity of transmission, where the difference d is also weighted by w. For example, if on average, descendants tend to have a smaller value of z than their parents, then d will be negative. Thus, all else equal this will tend to make the change in the average value of z negative (that is, the average value of z will decrease).

Again returning to our tea example, the preference for plain tea is transmitted with perfect fidelity, meaning that both parent and descendant have trait value of $z = 0$. This gives a difference of $d = 0 - 0 = 0$ between the trait value of plain-tea drinkers and that of their descendants. On the other hand, the preference for sweet tea is transmitted successfully only 2/3 of the time. The remainder of the time the descendant acquires a preference for plain tea. This means that the average trait value in the descendants of sweet-tea drinkers is $z = \frac{2}{3}1 + \frac{1}{3}0 = \frac{2}{3}$. And since sweet-tea drinking ancestors have a trait value of $z = 1$, the difference between their trait value and that of their descendants is, on average, $d = \frac{2}{3} - 1 = -\frac{1}{3}$. The value $E[wd]$ will therefore be negative. All else equal this therefore tends to decrease the average value of z (that is, the frequency of the preference for sweet tea decreases). Thus, the second term of Price's equation captures the effect of differences in the *fidelity* of reproduction among types.

Given our dissection of Price's equation it probably now comes as little surprise that this equation can be readily used to model the evolution of virtually any entity of interest, regardless of whether it is transmitted nongenetically or genetically. But Price's equation can also be used to model evolution when heredity is determined by a mixture of both genetic and nongenetic factors. It is this far-reaching generality that we, and others,[248] believe makes Price's famous equation perfect for formalizing a theory of evolution under extended heredity.

In the next chapter, we will make use of Price's equation to understand the potential evolutionary consequences of extended heredity. Of course, we are not suggesting that all evolutionary analyses must incorporate nongenetic inheritance. Although nongenetic inheritance is likely to be ubiquitous, the choice of modeling approach should depend on the specifics of the question of interest, and there may well be questions that can be addressed adequately using conventional evolutionary models. That said, even in such cases it can often be worthwhile to at least consider whether including nongenetic inheritance might provide a fresh perspective on the problem.

PRICE'S LEGACY

Although somewhat tangential to our purpose here, it is worth closing this chapter by briefly considering the significance of Price and his work in areas outside of extended heredity. Indeed, Price's equation has found great utility in many areas of evolutionary biology, including the evolution of altruism, multilevel selection, kin selection, and the so-called Fisher's fundamental theorem of natural selection to name a few. His equation manages to achieve all of this by reaching outside of the confines of population genetics and grasping more abstractly what all processes of selection have in common.

Despite this monumental achievement and the clarity that it brought to so many issues both inside and outside of evolutionary biology, Price himself appears never to have been fully satisfied with his accomplishments. Over time he grew increasingly depressed and, despite being a fervent advocate of science and a vocal critic of the supernatural in his early years, he took another sharp turn in the 1970s and converted to Christianity. He spent parts of the last years of his life obsessing over scriptures and at other times desperately trying to help the homeless of London. In one final twist of fate, Price eventually became homeless himself, and his life ended tragically by his own hand on January 6, 1975.

Although Price did not achieve widespread recognition for his work while alive, he is now recognized as having produced some of the most influential developments of evolutionary theory at that time. A growing number of articles and books now chronicle his remarkable life,[249] and his equation even inspired the 2006 feature film *The Killing Gene*, which stars the Swedish actor Stellan Skarsgård and a then little-known Tom Hardy.

7

Why Extended Heredity Matters

> No scientific theory is worth anything unless it enables us to predict something which is actually going on. Until that is done, theories are a mere game of words, and not such a good game as poetry.
>
> —J.B.S. Haldane, *Adventures of a Biologist*, 1937

In the preceding chapters, we have attempted to make the case for extended heredity in several ways. First, we detailed the inescapable logic that the universality of transmission of both genetic and nongenetic material from parent to offspring engenders a dual system of heredity. We then revisited the tangled history of how the current genocentric conception of heredity originated. We argued that the modern narrative, whereby a steady march of scientific progress ultimately disproved all other forms of heredity, is a somewhat revisionist portrayal of the history of the subject. Finally, we presented a survey of empirical evidence revealing that there is often much more to heredity than genes alone, and we developed a framework for how evolution should occur under an extended model of heredity.

In our view the collective weight of this logical, historical, and empirical evidence overwhelmingly demonstrates that genes should not be viewed as the only vehicles of information transfer between generations. Instead, the concept of heredity in evolutionary biology must be extended to include nongenetic material as well. Even so, this alone does not suffice to show that extended heredity is going to revolutionize our understanding of the evolutionary process. After all, a good deal of our current thinking about evolution was developed in Darwin's time, when the nature of heredity remained a mysterious black box, and "Lamarckian" phenomena such as the inheritance of acquired traits were accepted as fact.

The story of the full relevance of extended heredity in evolutionary biology remains to be written, as scientists continue to explore more deeply the diversity of nongenetic mechanisms of inheritance and to document the evolutionary implications of their findings. For now, however, enough initial research has been conducted that we can already begin to catch a glimpse of the outlines of such future developments. Our contention is that we can already see how our understanding of the evolutionary process, and our predictions for how evolutionary change will unfold under different conditions, is considerably altered by the presence of nongenetic inheritance. The purpose of this chapter (and the next two chapters) is to put flesh on the bones of this claim.

As a first step, we will explore a series of examples of increasing complexity, viewing evolution under extended heredity through the lens of the framework developed in chapter 6. Before doing so, however, we first take a brief detour into an idea from population genetics that will help us to visualize the evolutionary framework that we have developed.

A MOUNTAIN OF UNDERSTANDING

In the 1930s geneticist Sewall Wright developed an evocative metaphor referred to as the adaptive landscape to help visualize the evolutionary process.[250] The metaphor runs as follows. Picture the outline of a mountain along one dimension and think of the horizontal axis as representing the different possible states of a population. For instance, in terms of our example from chapter 6 involving tea preferences, the horizontal axis would represent the different possible frequencies of the preference for sweet tea in the population. Now imagine that the height of the mountain at any given point represents the average reproductive success of individuals in such a population. For example, the height of the landscape corresponding to a population containing a high frequency of sweet-tea drinkers will be larger than that of a population containing a low frequency of sweet-tea drinkers. This is because, as we assumed

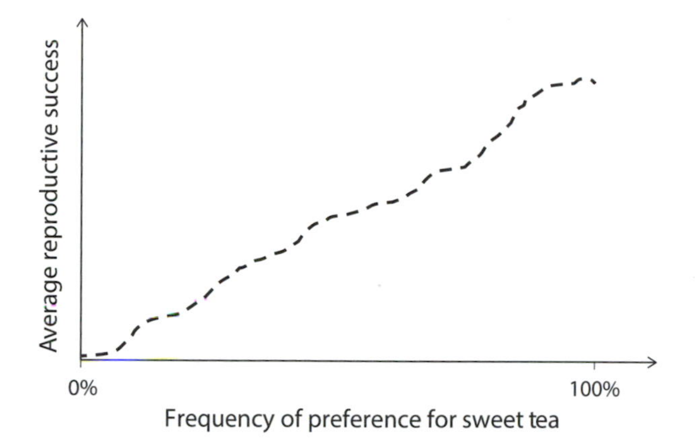

Figure 7.1. An adaptive landscape for the tea consumption example from chapter 6.

in chapter 6, those individuals with the preference for sweet tea have a greater reproductive success (that is, they teach more of the younger generation) than plain-tea drinkers (figure 7.1).

Now from our overview of Price's equation in chapter 6 we know that the first component of the equation (the covariance term) tends to drive the frequency of the trait in a direction of increased fitness. In other words, if having the trait (for example, the preference for sweet tea) causes an individual to have a high reproductive success, then the frequency of the trait will tend to increase. Conversely, if not having the trait (for example, not having a preference for sweet tea) causes an individual to have a high reproductive success, then the frequency of the trait would tend to decrease. Topographically, this therefore means that we tend toward a higher altitude on the adaptive landscape. The first component of Price's equation might thus be viewed as embodying an inexorable climb toward the summit of the mountain.

At the same time, the second component of Price's equation (the term $E[wd]$) also drives change in the frequency of the trait, and this change arises from a lack of fidelity during replication. In terms of our mountaineering metaphor this lack of fidelity means that some of the elevation gains achieved as we climb toward the peak are lost. The higher ground that is reached through the first term of Price's equation occasionally gives way beneath us, eroding our progress toward the summit. The highest point reached on the mountain, and how quickly we get

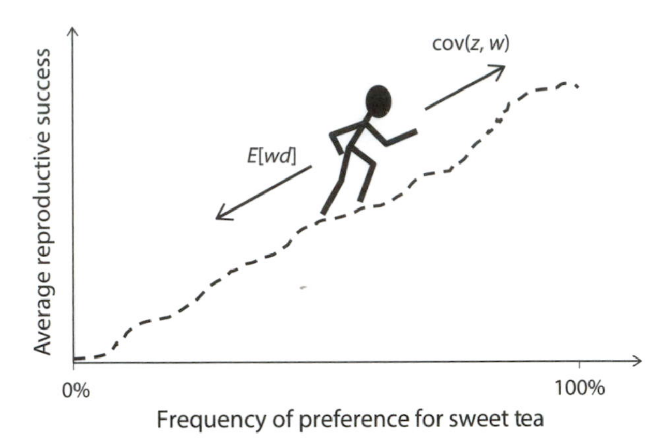

Figure 7.2. The relationship between the components of Price's equation and the adaptive landscape, in the context of the tea consumption example from chapter 6.

there, is therefore determined by a balance between these two processes (figure 7.2).

To see how we might use these ideas to understand the consequences of extended heredity, let's first consider a simple example comparing the evolutionary adaptation of two separate populations. In the first population, the trait of interest is transmitted entirely by nongenetic means while in the second it is transmitted entirely genetically.

In the first population, the trait is likely to flip back and forth between phenotypic states relatively quickly because epialleles and other nongenetic components of heredity are often relatively unstable. This heightened nongenetic mutation rate means that there will be a considerable amount of variation in the trait. Such high levels of variation will, all else equal, tend to make the first term of Price's equation relatively large. Thus our nongenetic population is quick out of the foothills, rapidly scrambling up the adaptive landscape. At the same time, however, this large mutation rate also means that the fidelity of transmission will be relatively low. This makes the second term of Price's equation relatively large and negative. Thus, before long our nongenetic population begins to lose its footing, ultimately reaching a statistical standstill part way toward the peak (figure 7.3a).

What about the genetic population? Genetic alleles are typically very stable, and this low mutation rate means that the trait variation in our

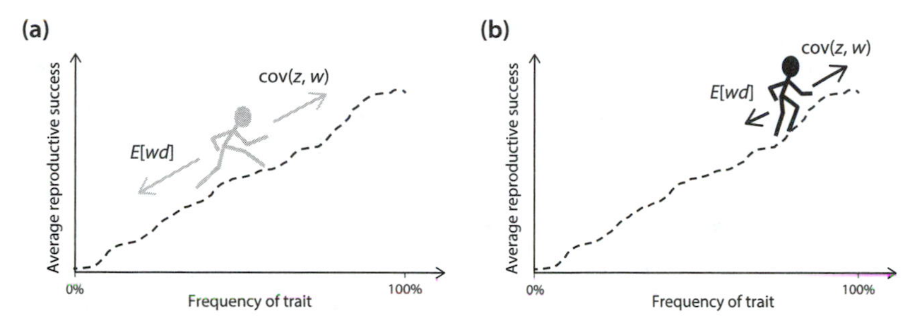

Figure 7.3. A pictorial depiction of the relative sizes of both components of Price's equation for nongenetic and genetic inheritance, as well as the consequences of these differences: (a) Evolution with nongenetic inheritance. (b) Evolution with genetic inheritance.

genetic population will be somewhat smaller than that of the nongenetic population. Thus the first term of Price's equation will be relatively small and the genetic population will be slow to get going. But the higher fidelity arising from a lower mutation rate pays dividends in the longer term because the population is able to maintain its footing throughout the climb. As a result, it ultimately reaches a higher altitude (figure 7.3b).

This evolutionary incarnation of the tortoise and the hare has been borne out in several theoretical studies.[251] The nongenetic hare makes initial progress toward the peak very quickly, but its progress is ultimately stymied. The genetic tortoise, on the other hand, slowly but surely reaches a higher ground. These studies have also shown that when both processes are allowed to occur within a single population, the population sometimes initially adapts to novel conditions very quickly through nongenetic means, but eventually the nongenetic underpinning of the trait gives way to a more stable genetic basis. Only when selective conditions continually fluctuate does nongenetic evolution continue to play a role in the longer term in these studies.

This example provides some simple insight into the role of extended heredity in evolution, but it is only the tip of the iceberg. For example, it assumes that nongenetic and genetic pathways of inheritance are independent of one another. This meant that we could consider the effects of each separately. But this is unlikely to be true most of the time. And when it isn't, we will not always be able to cleanly separate the evolutionary consequences of the two. Instead, we must employ two copies of

Price's equation, one copy for evolutionary change in the genetic component of the population and another for evolutionary change of the nongenetic component.[252] The first terms of each copy will be coupled to one another because reproductive success is often determined by a complex interaction between the genetic composition of an individual and the nongenetic milieu in which these genes find themselves. And the second terms of each copy will be coupled to one another as well because some heritable nongenetic material is often coded for by genes, and because patterns of gene expression (and even genetic mutation rates) can be affected by heritable nongenetic material.[253] We can begin to appreciate the implications of such interactions by revisiting the Venter experiment that we encountered in chapter 1.

THE VENTER CELL

The chimeric cell that Venter's group created out of a Frankenstein-like amalgam of a synthetic genome and a natural recipient cytoplasm provides a simplified opportunity to consider the evolutionary consequences of extended heredity when the two pathways of inheritance are not independent. As mentioned in the notes of chapter 1, Venter's group actually conducted two experiments. Both involved two closely related bacterial species of *Mycoplasma*, a genus that includes one of the causative agents of bacterial pneumonia in humans. The first species, *M. mycoides*, was designated as the genome "donor" and the second, a strain of *M. capricolum* colorfully named "California kid," was the genome "recipient."[254] Both species are opportunistic pathogens of ruminants including goats and sheep.

In the first experiment researchers extracted the genome from a donor cell and transplanted it into the California kid. Remarkably, this resulted in a living cell that, after several generations of replication, lost the phenotypic characteristics of the California kid and instead came to resemble the donor species. In a second experiment Venter's team aimed to take this feat of molecular biology one step further. Instead of using a natural donor genome in the transplantation, they constructed a synthetic copy of this genome from the ground up, using the chemical building blocks of DNA. This synthetic genome was then transplanted into the California kid as before. Now, however, despite the synthetic genome being identical in sequence to that of the natural genome used in the first experiment, the transplantation initially failed.

This failure is interesting because it demonstrates that DNA sequence information alone is insufficient, even to run a fully formed preexisting cell. Presumably something else, in addition to the DNA, was transplanted into the California kid in the first experiment when a naturally occurring genome was used. Unfazed, Venter's group pressed forward and eventually discovered the missing ingredient: DNA methylation. The natural genome of both the donor and the California kid are normally methylated, and this methylation appears to be necessary for the donor genome to function properly when transplanted. Once Venter's team methylated the synthetic genome in the appropriate way, the transplantation experiment worked largely as it had with a natural genome.

But does the DNA sequence information, along with its associated patterns of methylation, contain all the information needed to run a bacterial cell? While we don't know for sure, there are reasons to suspect not. First, another intriguing observation alluded to in Venter's study is that the reciprocal transplant, in which the genome of the California kid is transplanted into an *M. mycoides* cell, appears to pose difficulties.[255] Thus there seems to be a kind of interaction between cytoplasm and genome such that the combination of the two is not the sum of the parts. Second, only a tiny fraction of the transplants in the experiment worked. Of course, at one level it is remarkable that any of them worked, but it remains possible that in those instances where a successful transplant occurred, the cytoplasmic constitution of the California kid happened to more closely resemble that of the donor species in the first place.

It is also interesting to note that the phenotype of the chimeric cells was quantified and found to resemble that of the genome donor species only after three to ten days of growth. Given that cell replication occurs every eighty to one hundred minutes in this species, this means that upwards of fifty generations of cell division and evolution had passed before the analysis of the phenotypic patterns was possible. We can speculate about this period of cellular evolution in more depth using our theoretical framework for evolution under extended heredity.

During the fifty generations of cell division, both the genome and the nongenetic material were transmitted from parental cells to their offspring, with the latter including any epigenetic information and cellular constituents like proteins, RNA, and other biomolecules. During the lifetime of each cell, however, the composition of the nongenetic material was likely also supplemented with newly synthesized components that were encoded by the genome. Thus to track the evolutionary dynamics of the cell population during these fifty generations, we need to track the evolutionary change in the genome and in the nongenetic material as well as how the two interact.

We can extend our adaptive landscape metaphor to allow for this additional complexity. Both forms of inheritance, genetic and nongenetic, might be viewed as separate mountaineers on the landscape. Each might strive for their own peak individually, albeit perhaps at different rates, and each will also be subject to the hazards of loose footing, although perhaps to different degrees. But the interdependence of the two systems of inheritance means that our two mountaineers are also tethered to each other. The tether between them ensures that their evolutionary fates on the mountain are intertwined. In fact, although making for a slightly more comical metaphor, let's imagine that our climbers are connected to each other via a spring (rather like the ropes that real mountaineers use for safety, but more elastic). A soft spring means that each climber enjoys considerable independence of movement, whereas a stiff spring means that the evolutionary change in each of these two components of inheritance is rigidly coupled to the other.

Now in our example from Venter's study let's suppose that the phenotypic characteristics of a cell are entirely determined by the nongenetic material. For example, this might be true for the patterns of protein expression that Venter's group used as the phenotypic trait in their study. At the start of the experiment all cells had the California kid genome as well as the California kid nongenetic material, and thus they all had the California kid phenotype (figure 7.4).

These cells were then mixed into a solution containing the donor genome along with other chemical agents that facilitate the uptake of donor DNA by the California kid cells. The end result was a potpourri of California kid cells, some of which probably still contained the original California kid genome because they failed to take up the donor DNA, some that probably contained a mixture of the two genomes, and others that had their original genome entirely replaced by the donor. Overall,

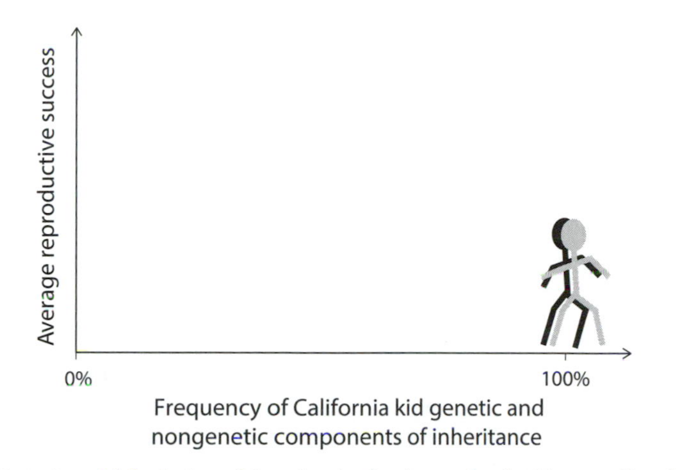

Figure 7.4. A pictorial depiction of the adaptive landscape for the Venter cell at the start of the experiment. Both the genetic and the nongenetic components of inheritance are characteristic of the California kid at the start of the experiment, and therefore the frequency of each component in the population is near 100%. Light gray indicates non-genetic material and black indicates genome.

the frequency of the California kid genome in the population was reduced in this solution through the transplantation process, but all cells still retained the California kid nongenetic material (and thus the California kid phenotype). This is illustrated in figure 7.5.

Now Venter's team was interested only in the cells that contained the donor genome and so, to separate the California kid cells with a donor genome from all of the rest, they employed a clever trick. When constructing the donor genome, they carefully inserted an extra gene coding for antibiotic resistance. When culturing the heterogeneous mixture of California kid cells that was produced by the transplantation process, they then laced the bacterial food with an antibiotic. This ensured that only those California kid cells with the donor genome would survive. This selection on antibiotic-laced food was continued for the fifty plus generations before the phenotype of the cells was measured.

The strong antibiotic-mediated selection in favor of the donor genome rapidly eliminated any cells containing only the California kid genome such that only cells containing the donor genome remained (and the *exclusive* presence of the donor genome was later verified by genome sequencing). Topographically, the antibiotic-laced food created a peak

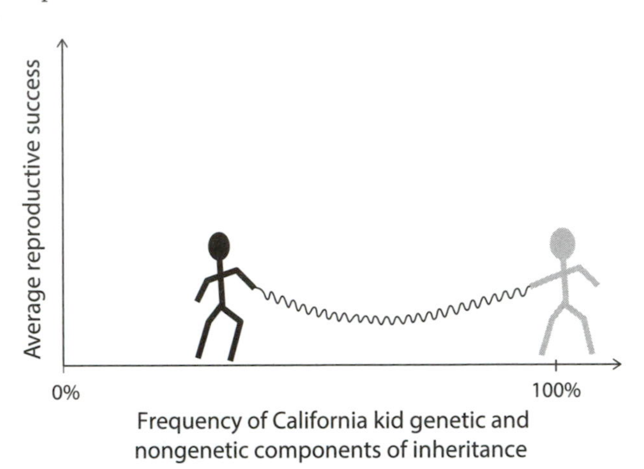

Figure 7.5. A pictorial depiction of the adaptive landscape for the Venter cell soon after the transplantation process. After genome transplantation, the frequency of the California kid genetic component of inheritance is reduced because many cells had their California kid genome replaced by the donor species' genome. The frequency of the California kid nongenetic component of inheritance, however, remains near 100%. Light gray indicates nongenetic material and black indicates genome.

on the adaptive landscape at the location where the California kid genome is absent, and our genetic mountaineer assuredly and powerfully scaled to the top of this peak (figure 7.6). From Price's perspective, the covariance term in the equation for genetic evolution was negative because cells without the California kid genome had the highest reproductive success when feeding on antibiotic-laced food. And the expectation term in the equation for genetic evolution was nearly zero because of the high fidelity of genetic transmission.

During the genomic ascent, the nongenetic material of the cells also changed. Presumably this nongenetic material did not have any independent effect on a cell's reproductive success, and therefore the covariance term of the equation for nongenetic evolution was zero—the nongenetic mountaineer made no attempt to climb the peak. But the fidelity of transmission of the California kid nongenetic material was not perfect. Some of the biochemical components of this material may have been self-regenerating (see chapter 2), and so they might have maintained their California kid identity for many generations. Others might have been more quickly dissipated through the dilution of cytoplasmic components that occurs every cell division, as the nongenetic

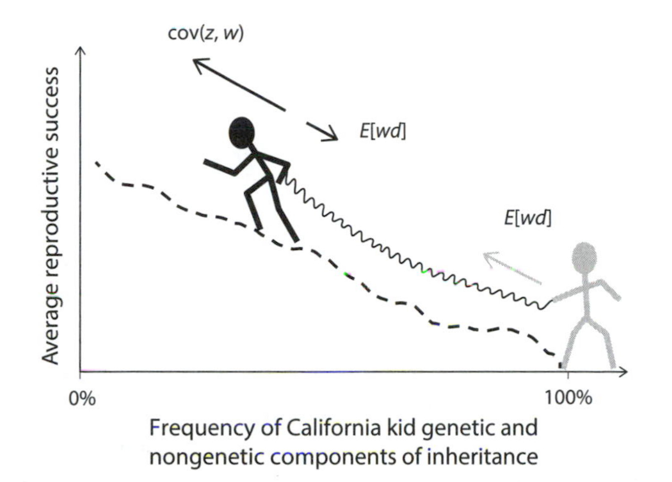

Figure 7.6. A pictorial depiction of the adaptive landscape for the Venter cell after antibiotic-laced food was introduced, along with an indication of the relative magnitude of each component of Price's equation. The presence of the antibiotic causes cells without the California kid genome to have the greatest reproductive success. This is indicated by the genetic adaptive landscape having a peak where the frequency of the California kid genome is 0%. Light gray indicates nongenetic material and black indicates genome.

composition of each cell was probably supplemented with newly synthesized constituents encoded by the genome. Thus, not only was the expectation term of the equation for nongenetic evolution negative (the California kid–type nongenetic material slowly morphed into a donor-type nongenetic material), but it was also coupled to the evolutionary dynamics of the genome. Our nongenetic mountaineer was therefore dragged to the summit of the mountain through the interaction between the two systems of inheritance. How quickly this occurred, and thus the number of generations of replication on the antibiotic-laced food that was required before all the cells of the population lost the features of the California kid, would be determined by the fidelity of its transmission (that is, the strength of the spring in figure 7.6). Nongenetic components that are strongly self-regenerating would take many generations to lose their California kid properties, whereas those components that act as passively inherited cytoplasmic elements would change more quickly.

How does this extended heredity view of evolution in Venter's experiment compare with what we might call the genes-only view of evolution? There are likely to be many different interpretations of what constitutes the genes-only view. Even so, it is probably fair to say that most adherents to a genes-only view of evolution would admit the existence of a dual system of inheritance in the case of bacterial cells, but they would claim that the genetic component of inheritance holds primacy over most properties of a cell. In effect, the process of DNA transcription during a cell's lifetime binds the cell's nongenetic material to its genome with a short and rigid spring. While this is probably true for some nongenetic components, those components that are self-regenerating (such as the proteins involved in self-sustaining loops of gene expression, discussed in chapter 2, or self-regenerating features of the cytoskeleton or cell membrane) can take a much more independent evolutionary path. And the more strongly self-regenerating they are, the more the predicted evolutionary outcome will be expected to differ from the genes-only view.

It is also instructive to speculate on how the experimental outcome might have differed under other conditions. For example, suppose that an antibiotic was not added to the food. In this case the covariance term in the equation for genetic evolution would be zero as well, and the adaptive landscape would be relatively flat. Our mountaineers would be thrown into the prairie-like world of figure 7.7, and the end result would likely be a mixture of cell types in the population.

Alternatively, suppose that we instead imposed selection for the California kid phenotype but that we did not put any antibiotic in the food. In this case cells with the California kid nongenetic material will have high reproductive success, and so the covariance term in the equation for nongenetic evolution would be positive rather than zero—our nongenetic mountaineer would now have a peak to climb. And so the expedition would unfold as shown in figure 7.8—the nongenetic material would evolve toward that of the California kid and the genome would follow. This is because the nongenetic material is continually supplemented with genome-encoded components, and those cells with a more California kid–like genome would have a higher reproductive success.

Again we might ask how this extended heredity view of evolution compares with the genes-only view. At one level, proponents of the genes-only view might rightly claim that this picture of evolution fits neatly within the traditional population-genetic framework—eventually

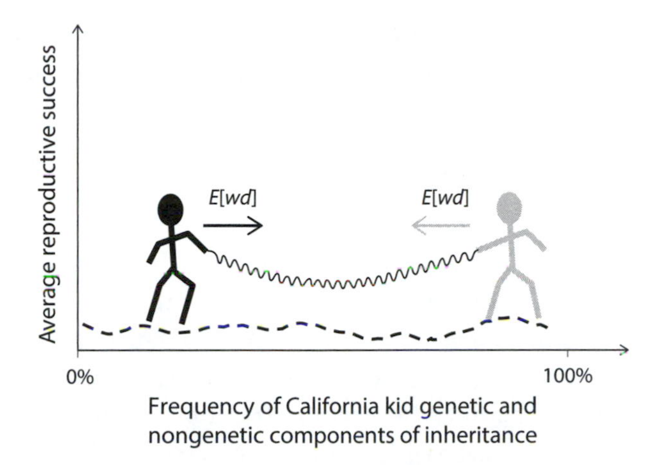

Figure 7.7. A pictorial depiction of the adaptive landscape for the Venter cell if antibiotic-laced food had not been introduced, along with an indication of the relative magnitude of each component of Price's equation. Neither the genetic nor the nongenetic component of inheritance has a fitness peak to climb. Light gray indicates nongenetic material and black indicates genome.

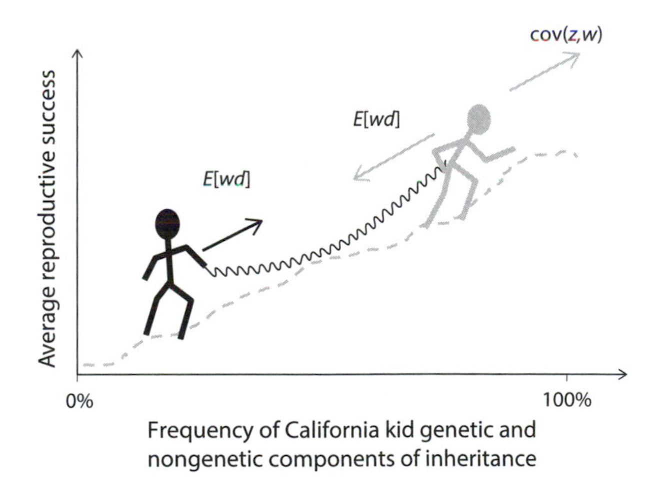

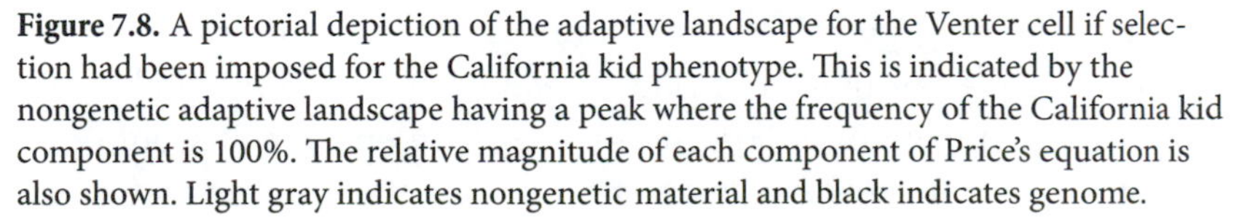

Figure 7.8. A pictorial depiction of the adaptive landscape for the Venter cell if selection had been imposed for the California kid phenotype. This is indicated by the nongenetic adaptive landscape having a peak where the frequency of the California kid component is 100%. The relative magnitude of each component of Price's equation is also shown. Light gray indicates nongenetic material and black indicates genome.

the population evolves to a point where only cells with both the California kid nongenetic material and genome are present. Furthermore, the genome would code for the adaptive California kid nongenetic material and so we would have a clear case of adaptive evolution of a genetically determined trait.

At another level, however, this claim would unfairly appropriate a subtle but important insight obtained from the extended heredity view—namely, that evolution actually occurred through selection and transmission of the nongenetic material. Although the genes-only view can readily allow for selection acting on the nongenetic material (that is, the phenotype) rather than the genome, traditionally this view assumes that the nongenetic material is reconstituted anew each generation by the transcription of the genome. As a result, any evolutionary change in the nongenetic material must necessarily be accompanied by an evolutionary change in the genome. But the extended heredity view allows us to see that this need not be true, and the more independent the two systems of inheritance are, the more freedom there is for evolution in the absence of genetic change.[256] Of course, one could readily fold this new insight stemming from extended heredity into the existing Modern Synthesis of evolution, but it seems disingenuous to claim that the Modern Synthesis *already* accounts for it. It seems clear that extended heredity does have the potential to offer fresh and novel insight into the evolutionary process and to generate predictions that deviate considerably from those of conventional population-genetic analyses. At the same time, the very fact that a single mathematical formalism can be used to describe both genetic and nongenetic evolution means that extended heredity does not really represent a radical departure from the fundamental principles of Darwinian evolutionary theory.

As one final thought-experiment, let's imagine instead that we had used antibiotic-laced food as done by Venter's group but that we also imposed selection for the California kid phenotype. In this case our mountaineers would each have their own peaks to climb. Our slow but sure-footed genetic climber will march steadily toward the donor genotype while our quick but clumsy nongenetic climber will be pulling in the opposite direction (figure 7.9). The end result would presumably depend on the relative strengths of the various factors involved. In this case we can see that predictions from our extended

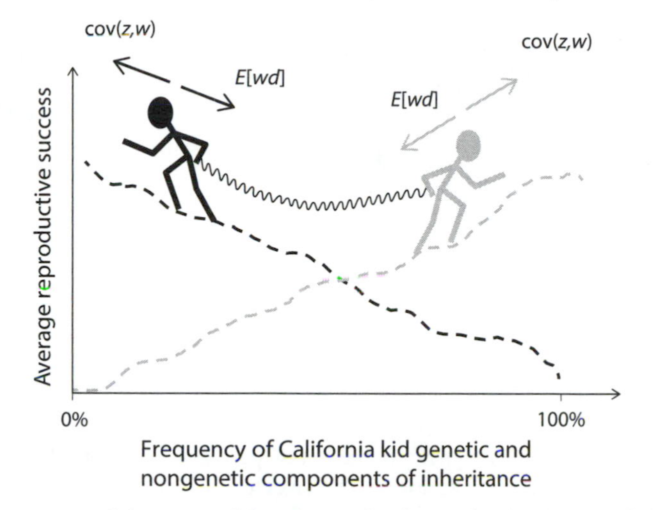

Figure 7.9. A pictorial depiction of the adaptive landscape for the Venter cell if selection had been imposed for the California kid phenotype and the donor species genotype, along with an indication of the relative magnitude of each component of Price's equation. Light gray indicates nongenetic material and black indicates genome.

heredity view of evolution can deviate even more substantially from the genocentric view.

GOT MILK?

These insights from bacterial evolution can also be applied in a completely different biological context—that of gene-culture coevolution—illustrating the unifying perspective provided by extended heredity. In many cultures milk and other dairy products like cheese and cream have become staple components of diet. Indeed, for some societies it is difficult to imagine how things could ever have been otherwise. Yet this is not the case for other mammals. Indeed, although milk production and consumption is a defining feature of this group of animals, it is extremely unusual for most mammals to continue consuming milk into adulthood.

Milk and other dairy products are rich in the sugar lactose, and the digestion of these food sources requires an enzyme called lactase. Lactase is produced in abundance in newborn mammals, but its production

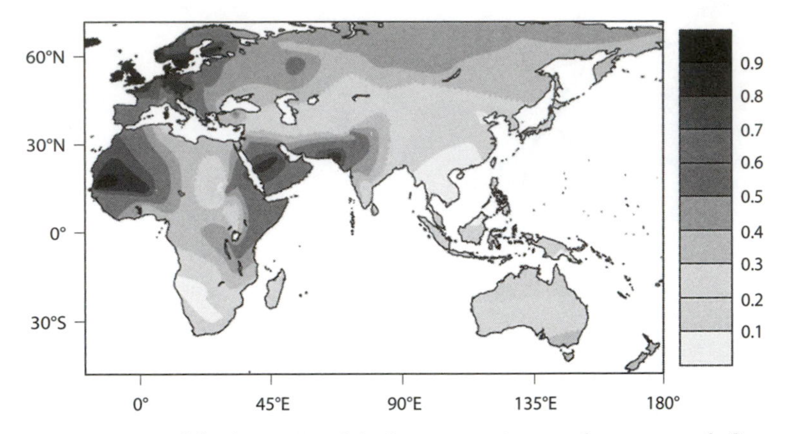

Figure 7.10. A map of the frequency of the lactase persistence phenotype excluding North and South America. (Redrawn from Itan et al., "A Worldwide Correlation of Lactase Persistence Phenotype and Genotypes," 2010.)

slows dramatically upon weaning. As a result, the consumption of dairy products after maturity typically results in nausea, bloating, and diarrhea. In humans, the phenomenon is referred to as lactose intolerance. In some human populations, however, an allele that causes the lactase enzyme to continue being produced throughout adulthood is very common. For example, this is true of many countries in northern Europe. This allele is called the lactase persistence gene, and individuals carrying a copy of this allele are able to consume and digest dairy products throughout their entire lifetime.

It was originally believed that lactase persistence was the "normal" state and that lactose intolerance represented a genetic defect. We now know, however, that the true story happened the other way around. Not only is lactose intolerance the ancestral state of human populations but even today the majority of humans are lactose intolerant (figure 7.10). The initial, erroneous, idea of lactose intolerance being a genetic defect stemmed from the fact that most of the initial research on the consumption of dairy products was conducted by northern Europeans, and northern Europeans tend to carry the lactase persistence allele. In hindsight, this Eurocentric view represents a clear example of how scientific research can be strongly influenced by the sociological setting in which it takes place.

Although people often speak of *the* lactase persistence allele, in fact, multiple different alleles that result in lactase persistence have arisen

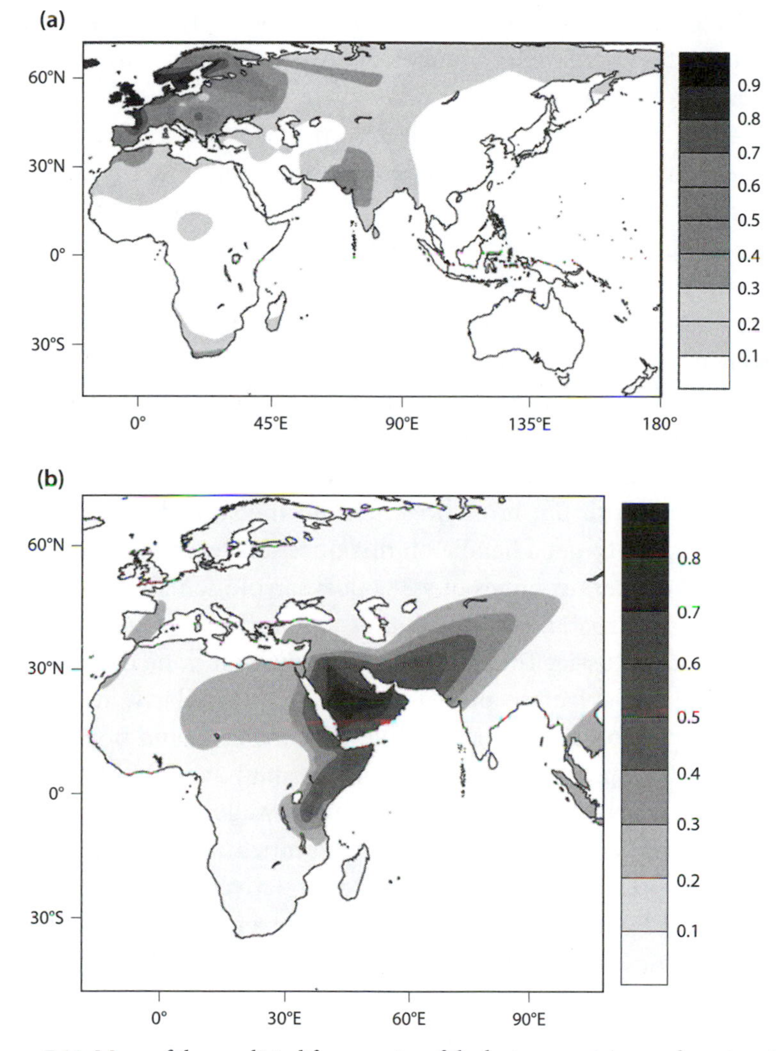

Figure 7.11. Maps of the predicted frequencies of the lactase persistence phenotype if it were determined by different genetic alleles: (a) Predicted frequency of the lactase persistence phenotype based on the 13910T allele. (b) Predicted frequency of the lactase persistence phenotype based on all known lactase persistence alleles except the 13910T allele. (Redrawn from Itan et al., "A Worldwide Correlation of Lactase Persistence Phenotype and Genotypes," 2010.)

in different parts of the world. One allele, labeled 13910T, is responsible for lactase persistence in much of northern Europe (figure 7.11a), whereas a combination of three other alleles, labeled 13907G, 13915G, and 14010C, are responsible for lactase persistence in Africa and the Middle East[257] (figure 7.11b). Notice, however, that there is still some discrepancy between the frequency of the lactase persistence phenotype shown in figure 7.10 at certain geographic locations and the frequency of these four specific alleles in figure 7.11, suggesting that either additional alleles for lactase persistence remain to be discovered, or that nongenetic factors also play a role.

The plurality of lactase persistence alleles and their independent origins and spread at different geographic locations suggests that the trait of lactase persistence was selectively advantageous in certain parts of the world. But how and when did the spread of lactase persistence occur? To get a handle on this question researchers extracted DNA from the femur bones of a 38,000-year-old female Neanderthal discovered in Croatia, and from other similar bones found in Spain, Germany, and Russia. They were able to show that none of these samples contained a lactase persistence allele.[258] Similarly, researchers investigated Neolithic and Mesolithic remains of *Homo sapiens* from between 4,000 and 6,000 years ago in Germany and Lithuania, and again an analysis of DNA failed to turn up any allele for lactase persistence.[259] Together these results provide compelling evidence that the origin and spread of lactase persistence is relatively recent, and that it coincides, at least approximately, with the advent and spread of dairy farming in these regions.

Of course, the temporal coincidence of the spread of lactase persistence and the spread of the practice of consuming dairy products into adulthood is probably not surprising since any lactase persistence allele would likely not have been advantageous in the absence of the consumption of dairy products. At the same time, the cultural practice of consuming dairy products would likely provide a reliable and renewable source of nutrition only if its practitioners also carried a lactase persistence allele. This example of gene-culture coevolution thus presents us with a chicken-and-egg problem. The cultural practice of consuming dairy products would not spread in the absence of an ability to digest lactose, yet the lactase persistence allele that confers an ability to digest lactose would not spread in the absence of the cultural practice of consuming dairy products. So how did things ever get off the ground?

While we can't know for sure, we can begin to explore this question using our theoretical framework from chapter 6. Again, there are two systems of inheritance, a genetic one that governs the transmission of the lactase persistence allele and a nongenetic, cultural one that governs the transmission of the practice of consuming dairy products into adulthood. Unlike our earlier example involving the Venter cell, these two systems of inheritance will now interact through the first terms of Price's equation. This is because the reproductive success of an individual carrying the lactase persistence allele will depend on whether the individual is a dairy consumer. And likewise, the reproductive success of a dairy consumer will depend on whether the individual carries the lactase persistence allele.

We can also attempt to extend our adaptive landscape metaphor to account for this complexity. Each mountaineer, genetic and cultural, will have its own landscape to climb, but now these landscapes themselves can change beneath each climber. In effect, the movement of one mountaineer causes a change in the topography experienced by the other. For example, if the genetic climber is near the place where the lactase persistence allele is absent, then the adaptive peak for the cultural climber will be located where dairy consumption is absent. Likewise, if the cultural climber is located where dairy consumption is absent, then the adaptive peak for the genetic climber will be located where the lactase persistence allele is absent (figure 7.12).

But now imagine that we were somehow able to move either the genetic climber or the cultural climber to the other extreme. For example, suppose we move the cultural climber to the bottom of its adaptive landscape and hold it there (figure 7.13a). The landscape of the genetic climber that is left behind would then reverse itself because a change in the position of the cultural climber affects how natural selection acts on the genetic climber. Thus the genetic climber would then find itself at the bottom of its adaptive landscape (figure 7.13b). And as our abandoned genetic climber begins to traverse the landscape to its new peak, the ground under our cultural climber would then begin to lift (figure 7.13c), eventually rising up until both climbers were located on a summit at a new position on the landscape (figure 7.13d).

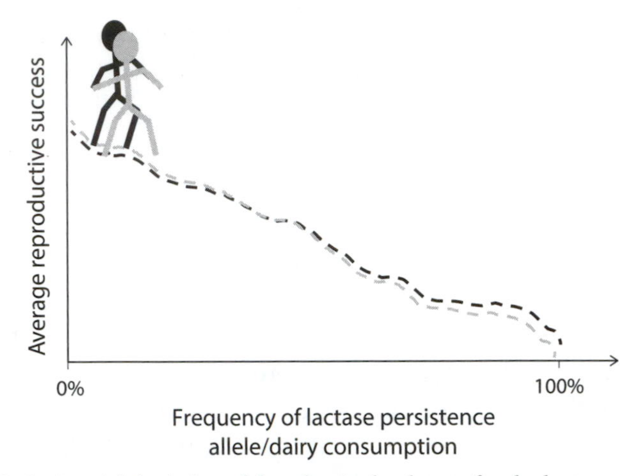

Figure 7.12. A pictorial depiction of the adaptive landscape for the lactase persistence allele and for the cultural practice of dairy consumption when both are absent. The adaptive landscapes for the lactase persistence allele and the cultural practice of dairy farming both have peaks at 0%. Light gray indicates cultural practice and black indicates genetic type.

From these considerations, it would seem that to get dairy consumption and lactase persistence both to spread, we somehow would need to perturb the ancestral population by either moving its genetic composition or its cultural practice wholesale to a new state and then holding it there until the other hereditary component evolved. And to make matters worse this move would decrease the fitness of individuals in the population. It is difficult to imagine how this might occur with the genetic composition of the population since it would require the simultaneous mutation of most individuals in the population from the lactose intolerant allele to the lactose tolerant allele. On the other hand, because some nongenetic factors, like cultural practices, can be induced to change by environmental conditions, perhaps this might provide an answer.

To see how this might work let's imagine an ancestral population of hunter-gatherer humans who lack the lactase persistence allele and who do not consume dairy. Indeed, it would be exceedingly difficult for people in such societies to obtain milk from wild animals. Sometime around ten thousand years ago, however, some human populations began transitioning to a more sedentary lifestyle, coupled with the advent of plant and animal domestication for grain and meat. This constituted a major

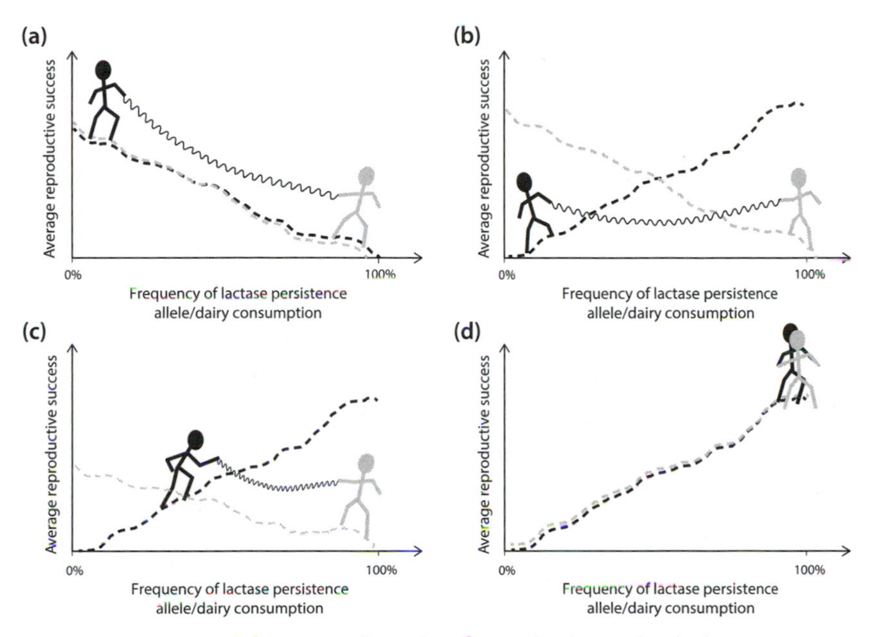

Figure 7.13. A pictorial depiction of how the adaptive landscape for the lactase persistence allele and for the cultural practice of dairy consumption might have changed over time during the evolutionary transition from a state where both were absent to the state where both are present. (a) The cultural climber is displaced to the bottom of its adaptive landscape from where it started in figure 7.12. This alters the landscape of the genetic climber such that it now finds itself at the bottom, as shown in (b). In (c) the genetic climber ascends its landscape. As it does so, the landscape of the cultural climber rises up underneath it, until the population reaches a state where the lactase persistence allele and dairy consumption are both present, as shown in (d). Light gray indicates cultural practice and black indicates genetic type.

change in the environment of such peoples since they would start to be in continued and close contact with domesticated animals. It is not hard to imagine that this environmental change led people to then experiment with using the milk from these newfound cohabitants as a reliable and renewable source of food, even though most such people would have been lactose intolerant. Put more abstractly, the "mutation rate" from the nonconsumer to the consumer state would have increased as a result of the environmental change. Such environmental induction, although maladaptive because most people could not digest dairy products, might still have resulted in the cultural practices of dairy farming and milk consumption throughout life then being passed on to subsequent generations. In this way, it might therefore have been the perturbation required to initiate the evolutionary transition from a population composed primarily of nonconsumer, lactose intolerant individuals to one composed primarily of dairy consuming, lactose tolerant people. In effect, environmental induction allowed the population to transition from one adaptive peak to another in a way that would have been extremely unlikely if inheritance was mediated by genes alone.

Whether this hypothesis is true or not remains to be determined, but it clearly shows how even malleable systems of inheritance with relatively low fidelity can help to shape the direction of evolution and to sustain complex, adaptive phenotypes over many generations. Furthermore, although this example involved cultural inheritance, in principle other nongenetic mechanisms could operate in exactly the same way (as we will see in chapter 9). Indeed, it is not difficult to imagine that there were other aspects of an individual's physiology and morphology that needed to be optimized for the consumption of dairy products, and that some of these were also heritable. For example, as we saw in chapter 1, some differences between Neanderthals and modern humans appear to result from differences in DNA methylation, and some of these patterns may be environmentally induced. Because we know that methylation patterns can sometimes be transmitted across generations independently of genetic alleles, these epialleles could thereby play a role in long-term evolution.

In this chapter, we have summarized some of the reasons why, in our view, extended heredity can contribute significantly to our understanding of evolution. But not everyone is convinced. In the next chapter, we will confront the major critiques of extended heredity, and offer our own perspective on this controversy.

8

Apples and Oranges?

> All this suggests that we must not discard a budding research programme simply because it has so far failed to overtake a powerful rival.
>
> —Imre Lakatos, *Methodology of Scientific Research Programmes*, 1978

Extended heredity is a controversial idea. Some biologists view it as a Pandora's box of woolly minded misconceptions whose advocates fail to understand or pay proper homage to the scientific advances made during the formulation of the Modern Synthesis. Others feel that extended heredity does not represent a significant departure from the ideas and practices of evolutionary research that have developed in the intervening decades, and therefore does not constitute a serious challenge to the status quo. In this chapter, we will examine the key arguments put forth against extended heredity, carefully assess the logic and evidence used to support both sides of the argument, and articulate our own position in this debate.

We will focus on four major critiques. First, skeptics have argued that extended heredity confounds fundamentally different things: heredity is about genes, while nongenetic factors are a downstream consequence of genes, and therefore do not represent independent hereditary factors.[260] In their view, including nongenetic mechanisms within the scope of heredity is a bad case of miscategorization, a sloppy mixing of apples and oranges. Second, some critics argue that nongenetic inheritance might well be real but it's not directly relevant to the study of evolution. This is because, even if nongenetic factors can be transmitted independently of genes, such factors are too unstable and too limited in their range of variation to play an evolutionary role. Third, it is often asserted that, even if nongenetic inheritance can sometimes influence evolution, such cases represent a minor wrinkle rather than a substantial conceptual challenge for evolutionary biology. According to this view, it may sometimes be useful to take nongenetic inheritance into account, but such cases do not fundamentally alter our understanding of how evolution works or how it should be studied. Fourth, there has been much criticism of claims that nongenetic inheritance generates adaptive ("directed") variation and can therefore be regarded as an independent driver of adaptive evolution. While our positions on these issues align in many ways with those of other proponents of extended heredity, we will also highlight some points of disagreement. In the final section of this chapter, we will outline what we see as the major hurdles that must be overcome to conclusively demonstrate an evolutionary role for nongenetic inheritance, and consider how this might be done.

CAN NONGENETIC FACTORS BE HEREDITARY?

An objection some critics raise is that the genome shapes the epigenome and phenotype, so only genes can be regarded as independent units of heritable variation. And, if nongenetic variation is a secondary, downstream consequence of genetic variation then, for all intents and purposes, only genes matter in evolution. This view reflects the traditional genotype/phenotype dichotomy, and the more recent formulation that sees genes as "replicators" and bodies as mere "vehicles" built by genes. According to this view, nongenetic effects of parents on their offspring are also appropriately viewed as elements of the phenotype—developmental switches that evolve via selection on genes and operate under genetic control. Such switches might extend plasticity across generations but, just like classic within-generation plasticity, they can be viewed as genetically based adaptations whose evolution can be fully understood within the classic Modern Synthesis framework of natural selection acting on genetic variation.

We believe that this view overlooks important parts of the picture. It is of course true that the genome shapes many aspects of development, and it is quite possible that the heritability of some traits is purely genetic. However, as we saw in chapters 4 and 5, it is also clear that many nongenetic factors are independent of DNA sequence variation and can function as independent hereditary units. Just like genetic alleles, such nongenetic factors can be transmitted across generations and can respond to natural selection. Such factors can constitute an important component of phenotypic variation and can enhance offspring-parent resemblance. This is obvious in relation to one form of nongenetic variation: culture. Cultural differences (such as differences in language, diet, or dress) between long-established human groups, or between individuals in culturally diverse societies, are transmitted from parents to their offspring, but such differences are not rooted in DNA sequence variation. The same is true of culture-like behavioral traditions in nonhumans, such as local song dialects in birds or tool kits in chimpanzees. We know this because such traditions can be transmitted "horizontally" between unrelated individuals, can change over an individual's lifetime, and can change across generations at a rate that far exceeds the plausible rate of genetic change (figure 8.1). Similarly, as we have seen, there is a great deal of evidence that some epigenetic, structural, and cytoplasmic variation can be acquired stochastically, induced through exposure to specific environments, or even generated experimentally, and this variation can then be transmitted to offspring and sometimes beyond.

Some of these effects are controlled by genetic switches that evolved to respond adaptively to the environment. For example, the induction of defensive spines in offspring of predator-exposed *Daphnia*, discussed in chapter 5, is a genetically based plastic response that regulates a developmental switch between two alternative phenotypic states. Although such parental effects can influence evolutionary dynamics, they could be regarded as fully genetically determined and therefore encompassed by conventional evolutionary theory. But a large component of nongenetically transmitted variation cannot be shoehorned into such a framework. After all, as we have seen, nongenetic inheritance allows for the transmission of a great variety of spontaneous, age-related, or environment-induced variants that result from changes in the epigenome, cytoplasm, soma, or behavior, and are mostly deleterious or neutral in their effects. Like genetic mutations, nongenetic variants such as the cortical abnormalities of single-celled eukaryotes, the epialleles

Figure 8.1. Culture can vary and change independently of genes. Although less familiar, other kinds of nongenetic factors (such as epialleles or structural variants) can also vary independently of genes and appear to be capable of similarly rapid and independent change. Such nongenetic factors can therefore be included within the scope of heritable variation. (© Gable/CartoonArts International)

that confer peloric flower shape in toadflax or the stress-induced behavioral syndromes of mice are transmitted to offspring as nonadaptive by-products of physiological and reproductive processes. Such factors contribute to heritable variation and form part of the "interpretive machinery" that regulates the expression of genes, but are not "genetically determined" in any meaningful sense.

CAN NONGENETIC INHERITANCE PLAY A ROLE IN ADAPTIVE EVOLUTION?

Even if nongenetic variation can arise and be inherited independently of genes, it doesn't follow that nongenetic inheritance plays a role in evolution. For one thing, critics have argued that nongenetic factors are just too unstable. If such factors change spontaneously or undergo environmental induction in every generation, then they might contribute nothing more than developmental noise. Even if transmission over several generations is possible, a high mutation rate would continually erode the phenotypic changes brought about by natural selection. We illustrated this problem in chapter 7 with the analogy of the nongenetic mountaineer whose advance toward the peak of the fitness mountain is impeded by the unstable ground underfoot, and formal analysis confirms that this is a real problem at plausible rates of epimutation.[261] Consequently, although persistent and intense selection on nongenetic factors could, in principle, push a phenotype toward the fitness peak, it's difficult to imagine how such nongenetic evolution could build a complex structure like the vertebrate eye, because even a relatively brief interlude of weak or reversed selection would destroy everything that had been built up over previous generations. In other words, although our nongenetic mountaineer might make it some of the way up the mountain by running hard, even a moment's rest will lead to a tumble down the slippery slope. Contrast this situation with genetic evolution. The stability of genes means that, even if selection is relaxed, the genomic machinery can persist with relatively little damage for many generations. For example, the ancestors of snakes appear to have been nearly blind but much of the genetic tool kit required for the development of eyes survived in their genome, giving their descendants the capacity to quickly reevolve decent eyesight.[262] Highly mutable nongenetic factors could not play such a role.

Yet, this critique overlooks other ways in which nongenetic inheritance can influence evolution. For one thing, as we noted in chapter 7, semistable nongenetic factors could play an important role in rapid evolution. High stochastic mutability and the potential for environmental induction will tend to generate a great deal of heritable nongenetic variation and, because a trait's capacity to respond to natural selection depends on the availability of heritable variation, theory suggests that the initial phase of adaptive evolution might often occur nongenetically.[263] That is, our fleet-footed nongenetic mountaineer will scale the fitness mountain first, leaving his genetic friend far behind. But of course there's a catch: like the tortoise who ultimately beats the hare to the finish line, the slower but more sure-footed genetic mountaineer may eventually ascend to an even higher fitness peak. In other words, if selection acts consistently over many generations, a nongenetically based phenotype may later be replaced by a more stable genetically based one.

The tortoise and hare analogy implies that nongenetic inheritance can play only a transient role, as an evolutionary mechanism of first response. But nongenetic inheritance could also play a long-term role, for a simple reason: many fitness peaks just don't stay put for long. In some contexts, such as coevolution between hosts and their parasites, the direction of natural selection is continually changing, forcing populations to chase an ever-shifting target. In such cases, the fleet-footed nongenetic mountaineer may be forever dashing hither and thither in pursuit of the elusive fitness peak, and his plodding genetic friend may never get a chance to catch up. While we do not yet know for sure what role nongenetic inheritance actually plays in such coevolutionary scenarios, we will see in the next chapter that it could be a key player.

But there is another important evolutionary role that can be played by even the most unstable types of nongenetic factors; they can interact with genes, resulting in powerful feedback loops that influence the course of genetic evolution. For example, cultural factors can interact with genetic factors, resulting in gene-culture coevolution. As we saw in chapter 7, such a feedback process almost certainly drove the evolution of the ability to digest cow's milk after weaning in some human populations that domesticated cattle and began to use unprocessed milk in the adult diet. As we will see in the next chapter, analogous feedbacks can occur in nonhuman and noncultural contexts as well. Moreover, even when nongenetic factors are fully determined by genes (as we assumed in some of our discussion of the Venter cell in chapter 7), or function as developmental on/off switches predictably triggered by parental environment, the presence of these nongenetic factors can still alter evolutionary dynamics and trajectories. Thus, interactions between genetic and nongenetic inheritance systems can lead to evolutionary outcomes that would be unlikely or even impossible with genetic inheritance alone.[264] Understanding and predicting such outcomes requires incorporating extended heredity into evolutionary theory.

A somewhat different critique of the potential for nongenetic inheritance to contribute to evolution is that only genes are capable of what David Haig has called "cumulative, open-ended change." David Haig, Doug Futuyma, and others have argued that nongenetic factors appear to be limited in their range of variation, typically alternating between just two possible states.[265] For example, epialleles are generally thought to turn a gene's expression on or off without altering the gene product produced. Likewise, *Daphnia* will produce offspring with or without spines depending on whether predator cues are present or absent, but exposure to a novel type of predator will not induce a new, qualitatively different response. Thus, while nongenetic traits are expected to be highly variable in that multiple phenotypes will be maintained in the population (for example, there will be individuals with and without spines), the range of possible phenotypes might still be quite limited. By contrast, DNA sequences are capable of almost unlimited variation because, even though each base can vary between only four possible states (A, T, G, or C), the long sequence of bases in a genome can be arranged in a vast number of possible combinations, corresponding to a potentially unlimited range of phenotypes. Genetic evolution therefore allows for the gradual buildup of complex adaptations, allowing a primitive chordate's light-sensitive eye spot to change through myriad evolutionary steps into the sophisticated vertebrate camera eye. Critics argue that selection on epigenetically controlled on/off states of existing genes could not produce such complex, novel adaptations. In other words, DNA sequences provide the hereditary underpinnings of evolution because their enormous combinatorial complexity allows for "unlimited heredity," while many nongenetic mechanisms of inheritance only allow for "limited heredity."[266]

While we do not deny that genes have a special role to play in long-term, cumulative adaptation, we do not believe that the evolutionary role of nongenetic factors is necessarily as limited as some critiques suggest. As Eva Jablonka and Marion Lamb have pointed out,[267] even if each epiallele (or other nongenetic factor) has just two possible states, the number of independently varying nongenetic factors is very large. And, if two systems both have nearly infinite degrees of freedom, it's a moot point that one has even more degrees of freedom than the other.[268] The range of variation that can be produced by the epigenome is large enough to generate the many vastly different cell types in the body of an animal or plant (although the range of epigenetic variation that can be transmitted through the germ line may be considerably smaller). Add to that the variation in other types of nongenetic factors, and the total range of heritable nongenetic variation—that is, the variety of possible phenotypes that can be produced through variation in heritable nongenetic factors—is surely very large.

The information content of many nongenetic factors is also enhanced by their nature. Although epialleles are typically regarded as regulatory switches that turn genes on or off, it's likely that variation in the degree of methylation of a gene's promoter or protein-coding regions, or subtle variation in chromatin structure, can allow for fine tuning of gene expression. Epigenetic mechanisms may also regulate alternative splicing of RNA transcripts and thereby shape the structure of proteins. Similarly, while parental effects are often described in terms of the presence or absence of a trait in offspring, most such effects can probably assume a range of values. For example, a parent's diet or behavior might influence the degree of preference or aversion that its offspring exhibit toward a given type of food, and parental nutrition probably has a quantitative effect on offspring growth. This reflects a basic difference between the way biological information is encoded in DNA sequences versus most nongenetic factors. As philosopher Peter Godfrey-Smith has emphasized, genetic information is stored in linear sequences of repeating units, whereas nongenetic hereditary information is largely analogue in nature.[269] In this sense, DNA sequences are like the digital information storage used by computers while nongenetic factors function more like the tuning pegs on a violin. Several pegs that can be tuned independently of one another (each capable of setting a string to any pitch within a certain range) can store far more information than an equivalent number of on/off switches.[270]

Of course, none of this refutes the fact that only genes appear to have the stability necessary for long-term, open-ended, cumulative evolution. Selection on genetic variation can produce adaptations as complex as eyes and brains, while selection on nongenetic factors is unlikely by itself to produce anything so impressive. But even if nongenetic factors cannot play precisely the same role as genes, they can play other important roles in evolution. Crucially, even highly unstable nongenetic factors can influence evolution because, as we will see in the next chapter, in the context of many evolutionary questions, the long-term stability of hereditary factors matters less than offspring-parent resemblance. And, of course, microevolutionary effects can set populations on new evolutionary paths and thereby have macroevolutionary consequences. Thus, both genetic and nongenetic inheritance systems are probably important in evolution, but their evolutionary roles are likely to be somewhat different.

DOES EXTENDED HEREDITY FUNDAMENTALLY CHALLENGE OUR UNDERSTANDING OF EVOLUTION?

Some critics also argue that, although nongenetic inheritance occurs and may even influence evolution in certain cases, its role is of insufficient general importance to speak of a fundamental challenge to established evolutionary theory or the conventional practices of evolutionary research. According to this view, the standard assumptions may be simplifications of the messy real world, but these assumptions are the basis of a powerful and elegant paradigm that approximates reality very well in most cases.

But the claim that extended heredity represents a trivial extension of established theory is difficult to square with the view, expressed by many prominent evolutionary biologists, that extended heredity violates assumptions central to Modern Synthesis theory—in particular, the assumptions that genes are the sole basis of heredity, and that environmentally induced ("acquired") traits cannot be transmitted to descendants.[271] As we saw in chapter 3, architects of the Modern Synthesis like T. H. Morgan, Julian Huxley, Theodosius Dobzhansky, and Ernst Mayr fought tooth and nail against any suggestion that these assumptions might be violated, and consistently singled out "Lamarckian inheritance" as the quintessential evolutionary heresy. John Maynard Smith called nongenetic inheritance "the only significant threat to our views."[272] The attitude of these leading evolutionary biologists speaks to the centrality of the exclusively genetic concept of heredity to established evolutionary theory. Although evolutionary biology has acquired many new tools and ideas in the decades since the Modern Synthesis was developed, the same assumptions still underpin much of evolutionary research today.

Biologists' reluctance to embrace extended heredity may stem in part from the fact that it complicates both theoretical and empirical research. Population geneticists conventionally model evolution based on a simple segregation of alleles in accordance with Mendelian rules. As we showed in chapters 6 and 7, extended heredity complicates these models by adding at least one inheritance channel that operates under a different (and often poorly understood) set of rules. Even greater complications arise in empirical studies. Quantitative genetic analysis is based on the fundamental assumption that, after controlling for common environment and maternal effects, any remaining resemblance between relatives must have a genetic basis.[273] Extended heredity can violate this assumption because, with nongenetic inheritance, phenotypic similarity between relatives such as paternal half-siblings can have either genetic or nongenetic causes. This can lead to inflated estimates of parameters such as additive genetic variance and heritability, because these values can reflect a combination of genetic and nongenetic effects. Even the observation that identical twins tend to be more similar than fraternal twins in some traits cannot necessarily be used as evidence of a genetic basis for those traits because identical twins arise from the same egg and sperm and could therefore share cytoplasmic and epigenetic factors that are not shared by fraternal twins.[274] Likewise, genome-wide association studies (GWAS) search for associations between DNA sequence variants and phenotypic traits, and therefore cannot detect nongenetic causes of variation in phenotype. This could lead to paradoxical patterns like "missing heritability."[275] From a practical perspective, extended heredity is a can of worms, and some scientists understandably dread the surprises that lurk under the lid. But, having discovered that there are more things in heredity than are dreamt of in classical genetics, we have no choice but to let the worms crawl where they may.

A different objection is that nongenetic inheritance is just the tip of a very big iceberg—the role of environmental effects on development (that is, developmental plasticity) in evolution. Mary Jane West-Eberhard has argued that classic within-generation plasticity is a far more widespread, well-documented, and important phenomenon than nongenetic inheritance, so the excitement over nongenetic inheritance is misplaced.[276] Yet, the number and variety of examples of nongenetic inheritance is growing rapidly, so much so that it's no longer obvious that nongenetic inheritance is much less widespread than within-generation plasticity. Moreover, while West-Eberhard and others have argued persuasively that plasticity's role in evolution deserves greater attention, there are equally good reasons to believe that nongenetic inheritance can lead to interesting and unexpected evolutionary outcomes. But, while biologists have never denied the direct role of environment in shaping development (although they may have tended to overlook its importance), the very existence of nongenetic inheritance was denied for many years, presenting today's biologists with the challenge of understanding this long-neglected phenomenon.

CAN NONGENETIC INHERITANCE DRIVE ADAPTIVE EVOLUTION WITHOUT NATURAL SELECTION?

Most controversial of all has been the claim that nongenetic inheritance allows organisms exposed to a novel environment to acquire and transmit those particular features that will enhance the fitness of their offspring—that is, that nongenetic inheritance disproportionately generates adaptive "directed variation." For example, a recent paper argued that "heritable variation will be systematically biased towards variants that are *adaptive*."[277] Proponents of this idea even claim that, by generating such directed variation, nongenetic inheritance becomes a mechanism of adaptive evolution in its own right, capable of generating adaptive change without the help of natural selection. Eva Jablonka and Marion Lamb referred to the supposed tendency of nongenetic inheritance mechanisms to generate adaptive variation as an "instructive" process and suggested that "evolutionary change can result from instruction as well as selection."[278]

Such suggestions have been particularly unpalatable to critics[279] because they challenge Darwin's most important insight—the idea that adaptation results from natural selection on random variation.[280] To better understand the debate we need to consider what "random variation" actually means.

Consider a population of individuals exposed to one of two different *evolutionarily novel* environments, "cold" or "hot" (figure 8.2). By evolutionarily novel, we mean that the lineage has not previously encountered and adapted to those environments in the course of its evolutionary history.

The conventional view of random variation is that, if we consider the pool of mutational variants produced by all individuals in the population, then this pool will contain the same relative abundance of each

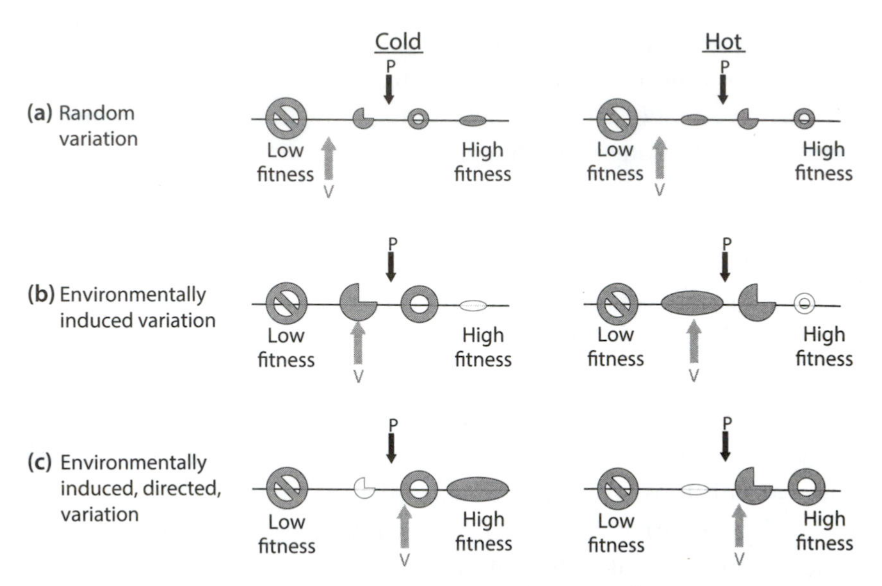

Figure 8.2. Each panel (a)–(c) shows a different type of mutational process in two hypothetical environments, "Cold" or "Hot." Shapes represent different types of mutant alleles, or epialleles or other nongenetic variants produced in the population in each environment. Size of shape indicates its relative abundance in the pool of newly produced variants in the population as a whole. All variants are ordered left to right with respect to their fitness in the specified environment. Gray arrow labeled "V" is the average fitness of the variants that are produced. It can be thought of geometrically as the "center of mass" of all the variants along the fitness axis. Black arrow labeled "P" is the average fitness of the population that produced the variants. (a) Random variation—most mutations are deleterious regardless of environment. This is indicated by the average fitness of the variants "V" being less than the average fitness of the population that produced them, "P." The fitness of variants depends on the environment (reflected by their ordering being different in the two environments), but the relative abundance of each of type among the newly produced variants is the same in both environments. (b) Environmentally induced variation—again, most mutations are deleterious regardless of environment. This is indicated by the average fitness of the variants "V" being less than the average fitness of the population that produced them, "P." The fitness of variants depends on the environment (reflected by their ordering being different in the two environments), and the relative abundance of each type in the pool of newly produced variants differs between the two environments (e.g., the "ellipse" variant is not produced in the cold environment whereas the "doughnut" variant is not produced in the hot environment). Critically, however, the average fitness of the pool of all variants is still less than the average fitness of the population that produced them. (c) Environmentally induced, directed, variation—most mutations are advantageous regardless of environment. As in panel (b) the fitness of variants depends on the environment (reflected by their ordering being different in the two environments), and the relative abundance of each type in the pool of newly produced variants differs between the two environments (e.g., the "3/4 disk" variant is not produced in the cold environment whereas the "ellipse" variant is not produced in the hot environment). Critically, however, the average fitness of the variants "V" is greater than the average fitness of the population that produced them, "P."

variant regardless of whether the population is in the cold or the hot environment (figure 8.2a). Mutation is random in the sense that the environment has no effect on which variants appear. Typically, we also expect that the majority of variants produced will be deleterious, meaning that the average fitness of the variants will be less than the average fitness of the population that produced them. The Modern Synthesis holds that adaptation occurs as a result of natural selection acting on such random (genetic) variation. With environmentally induced variation the types of variants that occur tend to differ between the environments (figure 8.2b). This is the kind of variation that's often generated by developmental plasticity. From the standpoint of fitness, however, again we expect that the majority of variants produced will be deleterious. It is this form of variation that seems to occur quite frequently in nongenetic inheritance. Notice, though, that, in principle, there is no reason why genetic mutation couldn't also be environmentally induced as in figure 8.2b (indeed, as we have already mentioned, there is evidence of this kind of genetic mutation), but most contemporary evolutionary theory does not incorporate this possibility. Finally, with environmentally induced, directed variation the types of variants that occur tend to differ between the environments *and those that occur in each environment tend to yield high fitness in that environment* (figure 8.2c). This means that the average fitness of the variants will be larger than the average fitness of the population that produced them, and this will be true in both environments. It is this latter form of variation that has generated the most controversy because, if it were possible, it would provide another mechanism of adaptive evolution in addition to natural selection.

Another way to think about these different kinds of mutational processes is through Price's equation from chapter 6. Recall that Price's equation tracks evolutionary change in the average value of a trait from one generation to the next. If we take the trait of interest to be an individual's fitness, w, then the equation takes on a particularly simple form, telling us how the average fitness in the population changes over one generation. We get

$$\Delta \bar{w} = \mathrm{cov}(w, w) + E[wd] \tag{1}$$

Remember that the first term of Price's equation embodies the effect of natural selection. In Equation (1) this term is the covariance of fitness with itself, which is just the variance in fitness. Now the variance of any variable is positive (unless the variable is a constant, in which case it is zero). Therefore, the first term in Equation (1) will typically be positive, meaning that natural selection drives an increase in average fitness—the population climbs the adaptive landscape. The second term in Equation (1) is the change in average fitness as a result of mutation. In both figure 8.2a and figure 8.2b, this change is negative because most variants produced are deleterious. As a result, only natural selection causes population adaptation—at equilibrium natural selection driving adaptation will be balanced by deleterious mutation. With directed variation (figure 8.2c), however, the second term in Equation (1) will be positive because most variants produced are beneficial. This means that population adaptation occurs both through natural selection and through directed variation.

But how could directed variation occur? As we saw in previous chapters, some instances of nongenetic inheritance—that is, adaptive parental effects—have clearly evolved to enhance fitness.[281] For example, in variable environments, several nongenetic components of a parent's state (for example, epigenetic factors or resources) change plastically in a way that tends to enhance fitness (and we can think of these changes as a form of nongenetic "mutation"). Some of these components are then transmitted to offspring and thereby alter offspring development in a way that better suits the offspring to the anticipated environmental conditions. Such "anticipatory" parental effects clearly differ from random variation because their effects on offspring fitness are positive on average—in fact, anticipatory parental effects might produce a pattern identical to that depicted in figure 8.1c.

However, although adaptive parental effects can result in the transmission of factors that enhance offspring fitness in response to environmental challenges, we do not believe that such effects should be labeled a form of directed variation. Rather, like adaptive within-generation plasticity, adaptive parental effects are evolved mechanisms that allow organisms to respond adaptively to *evolutionarily familiar* challenges—that is, to challenges that are similar to those experienced by the lineage over many generations and to which the lineage has adapted through the evolution of a suite of fitness-enhancing responses. The distinction between adaptive parental effects and true directed variation therefore hinges on whether or not the environmental challenge is evolutionarily familiar or evolutionarily novel. In our view only if organisms can adjust the variation produced so as to enhance fitness in response to truly novel challenges can directed variation be said to occur.[282]

Do organisms possess the capacity to produce adaptive variation in response to truly novel circumstances and then pass these "mutations" on to their offspring? We believe that the answer is "yes," but only in a very limited sense. Human beings and, to a lesser extent, other cognitively sophisticated animals, have the capacity to find solutions to novel problems posed by their environment and can sometimes transmit such innovations to their offspring; this is the function of cognition, behavioral plasticity, and learning. But the scope for such cognitive innovations to drive adaptive evolution seems quite circumscribed—they can occur only in the most cognitively complex of animals, are likely to generate only short-term solutions, and are probably possible for a very restricted subset of the challenges that the world can present.

There can be little doubt that cultural evolution has transformed our species in profound ways, and it's possible that the collective intelligence of billions of people, powered by science and linked through the internet, will allow *Homo sapiens* to overcome even deeper challenges in the future. Yet, even human intelligence is notoriously poor at anticipating the long-term consequences of current actions, and what seems like a great idea today often proves disastrous in the long run (think fossil fuels, fast food, or nuclear weapons). Perhaps, as historian Yuval Noah Harari believes,[283] we are on the cusp of a new era of self-guided evolution enabled by genetic engineering technologies and driven by the desire to "improve" our bodies and minds. But whatever sorts of "designer babies" humans choose to create, the long-term outcome is not likely to be adaptive in any conventional sense. The evolution of *Homo sapiens* driven by the whims of its own brain will be a process unlike anything seen before in the history of life on earth, and we cannot see a plausible analogue to such a process in the evolution of other species. A bird may discover a clever way to open milk bottles to get at the milk fat under the lid, and this behavior might spread through social learning and lead to improved nutrition. But there is nothing to prevent the same birds from learning to access poisoned bait, and this maladaptive behavior might spread just as readily via social learning if the poison's effects are latent and cumulative, like smoking.[284] If the former behavioral innovation is preserved over multiple generations while the latter disappears, it will be natural selection that's responsible. Consequently, it's not clear how often behavioral plasticity actually leads to adaptive outcomes in truly novel circumstances,[285] and it's unlikely that such cognitive innovations can drive long-term adaptive change without the help of natural selection.

But most organisms lack brains, and can respond to their environments only through physiological changes. When it comes to such noncognitive forms of plasticity, the difficulties of explaining directed, adaptive variation are much greater still. Evolved mechanisms of plasticity will tend to enhance fitness in response to evolutionarily familiar environments and challenges. They may also enhance fitness under conditions that represent a modest extension of an evolutionarily familiar environmental gradient. For example, an evolved mechanism of developmental plasticity that allows organisms to cope with temperatures in the range of 15 to 30°C might also produce a reasonably well-functioning phenotype if the temperature rises to 35°C. Likewise, if temperature fluctuations often span multiple generations, the temperature parents experience may induce adaptive changes in their offspring through an evolved mechanism of transgenerational plasticity. But what if the environment becomes polluted with artificial hormone-mimicking chemicals? We can see no reason to believe that organisms would tend to respond adaptively to such an evolutionarily novel challenge, nor that they would spontaneously produce offspring that differed from their parents in ways that preadapted them to the new challenge. Rather, we would expect offspring to exhibit random genetic and nongenetic variation with respect to the new needs (for example, both greater and lesser sensitivity to the hormone mimics), and that adaptive plasticity and adaptive parental effects might evolve over many generations by natural selection on this heritable variation. In other words, in circumstances that are truly evolutionarily novel, nongenetic inheritance is likely to be random in its effects on fitness, just as is genetic inheritance.

A more modest form of the directed variation argument is that organisms have evolved to respond to novel challenges by preferentially generating mutations in those particular traits that play the most direct roles in dealing with that type of challenge. For example, Jablonka and Lamb point to studies suggesting that some bacteria have evolved the capacity to respond to changed nutritional conditions by preferentially mutating genes involved in the relevant metabolic pathways. Although many such mutants will still perish, "focusing" mutation on the most relevant genes means that the probability of hitting upon a lucky "solution" to the challenge is far greater than it would be if the mutation rate increased throughout the genome. Jablonka and Lamb argue that many organisms may employ analogous strategies of generating increased genetic and nongenetic mutation under stress, either overall or in the most relevant traits. Indeed, epigenetic factors such as DNA methylation can influence rates of genetic mutation at specific genomic regions, providing a potential mechanism for such responses.[286] A related idea is that evolution can "learn," meaning that natural selection hones lineages to be more evolvable and more likely to undergo adaptive transitions.[287]

It's certainly possible to imagine organisms evolving mutational mechanisms or analogous nongenetic variance-generating mechanisms for dealing with particular types of environmental challenges, but we see these as adaptations to specific challenges rather than as manifestations of a general tendency to generate adaptive mutations under stress. Some bacteria may respond to nutrient limitation by ramping up the mutation rate of genes involved in metabolism because nutrient limitation is a challenge consistently faced by bacteria over billions of years. But could bacteria possess analogous mutational mechanisms for every challenge that they could possibly encounter? This would require either a generalized mechanism enabling bacteria to recognize the cause of the stress that they are experiencing and identify the particular biochemical pathways that are especially salient to "solving" that problem or, alternatively, numerous specific mechanisms geared to dealing with a vast array of specific stressors. Both possibilities seem implausible. In fact, such mechanisms could exist only if no environment or challenge were truly evolutionarily novel—that is, if we were to assume that organisms had already experienced all contingencies that they could possibly encounter. Furthermore, beyond merely experiencing these challenges, organisms would have had to evolve cellular and physiological mechanisms that could respond adaptively to this multitude of potential challenges, that is, a vastly versatile form of plasticity that could be maintained over many generations and respond adaptively even to challenges that are encountered very rarely. It's not clear to us how such a mechanism or set of mechanisms could evolve or persist in lineages. In short, we remain to be convinced that evolution could endow organisms with a general capacity to optimize the phenotypes of their descendants for evolutionarily novel environments, thereby allowing for adaptation without natural selection.

THE MISSING PIECES OF THE PUZZLE

The existence of nongenetic inheritance is no longer in doubt. Although particular examples might ultimately prove to be illusory, the reality of parental effects and the processes of structural, cytoplasmic, epigenetic, symbiotic, and behavioral/cultural inheritance cannot be denied. But, from an evolutionary perspective, demonstrating that nongenetic inheritance occurs is only the first step. What would it take to demonstrate beyond any doubt that nongenetic inheritance plays a role in adaptive evolution?

The role of nongenetic inheritance is well established in the context of culture and human evolution. For example, as we have seen, the interaction of genes and culture almost certainly drove the genetic evolution of the lactase persistence allele and the cultural evolution of milk use in the adult diet in some human populations. However, while this and other examples of gene-culture coevolution provide proof of principle, these examples do not establish that nongenetic inheritance influences evolution in species other than *Homo sapiens*, or through mechanisms other than culture. To date, well-established examples outside the human and cultural context are still lacking.

To sharpen our focus let's consider the three ingredients that are required for adaptive evolutionary change in any trait to occur by natural selection:

1. The trait must vary among individuals
2. The trait must be heritable
3. The trait must affect the survival and/or reproductive success (that is, fitness) of individuals

Much of the past fifty years of success in evolutionary research has come from documenting these three ingredients, and the adaptive evolution that takes place for traits having a genetic basis, both in the lab and in natural populations. One might therefore consider using a similar approach for traits having a nongenetic basis.

From the research reviewed in previous chapters it is clear that evidence exists for all three of these ingredients in the context of nongenetic inheritance. Traits affected by nongenetic factors certainly vary among individuals, there is evidence for the heritability of these traits through various mechanisms of nongenetic inheritance, and the traits in question often affect individual fitness. Unfortunately, all three ingredients are not always documented within the same organism, but there are notable exceptions. For example, Frank Johannes and colleagues, working in the lab of Vincent Colot at the Centre National de la Recherche Scientifique (CNRS) in France, used the plant *Arabidopsis thaliana* to create a series of "epigenetic recombinant inbred lines" (epiRILs)—*Arabidopsis* strains that are nearly identical genetically but highly divergent epigenetically. They accomplished this by crossing two *Arabidopsis* strains that differed only in that one strain carried mutant copies of a gene involved in the maintenance of DNA methylation, and therefore had reduced DNA methylation throughout its genome. After crossing the two strains, they then back-crossed one of the resulting offspring to the normal strain and selected a series of descendants that were similar genetically but varied markedly in their patterns of DNA methylation. They were able to show that plants with different patterns of methylation differ in traits like flowering time and plant height (ingredient 1) and that the patterns of methylation underlying these differences are transmitted stably from parent to offspring, sometimes over many generations (ingredient 2).[288] Finally, we know from other studies that traits like flowering time and plant height are important determinants of fitness (ingredient 3). What remains to be seen, however, is whether these three ingredients would then combine to result in adaptive evolution in these traits if allowed to do so.

One way to begin addressing this question is to conduct artificial selection experiments in the lab. For example, one could artificially select for different plant heights or flowering times using the epiRILs and then see how much evolutionary adaptation occurs. As we saw in chapter 4, a few such studies on isogenic or highly inbred lines of mice, flies, and worms have already been conducted and do show that epigenetic traits can respond to natural selection.[289] Even in these cases though, it remains unclear whether the adaptation that occurred was entirely underpinned by nongenetic inheritance. In fact, epiRILs do not entirely circumvent the pitfalls inherent in research on epigenetic inheritance, because one of the functions of DNA methylation is the suppression of transposable elements—"parasitic" DNA sequences that can insert new copies of themselves throughout the genome, disrupting the activity of important genes. Because the epiRILs are derived from an ancestor that could not maintain normal levels of DNA methylation, they inherited methylation-deficient genomic regions that are susceptible to transposable elements, and so may harbor small but potentially important genetic differences as well.

At the same time, it might be naive to think that the consequences of genetic and nongenetic inheritance systems can be neatly separated. As we suggested in chapter 7, it is possible (perhaps probable) that the two systems interact in ways that make their combined effects completely different from what we might predict based on their individual effects. The emergence of new molecular technologies may offer researchers a way to begin addressing these issues by allowing them to manipulate specific epigenetic factors directly—for example, to delete or create particular DNA methylation patterns and to inhibit or induce specific noncoding RNAs. It is already possible to knock out or modify the DNA methyltransferase (DNMT) enzyme systems that maintain DNA methylation states, altering methylation throughout the genome.[290] It will probably soon be possible to modify the methylation states of particular genes[291] or selectively knock out certain noncoding RNAs.[292] This would allow researchers to directly establish the hereditary role of epigenetic factors such as methylation patterns and noncoding RNAs, independent of any genetic changes.

Beyond artificial selection experiments in the lab,[293] it will also be necessary to look for evolutionary adaptation via nongenetic inheritance in natural populations. Of course, the difficulties in doing so are further compounded beyond lab studies, and to date there is very little evidence along these lines. There are a growing number of studies that quantify epigenetic variation (like patterns of methylation) across natural populations. Many of these studies also compare the amount of epigenetic divergence across populations with that of genetic divergence and/or examine whether epigenetic patterns across populations are correlated with different environmental conditions or selective regimes.[294] As we will see in the next chapter, often the degree of epigenetic divergence exceeds that of genetic divergence, and epigenetic patterns also tend to correlate well with environmental conditions and/or traits thought to be differentially selected in the different populations. This has sometimes then been taken as supporting the possibility of adaptation via epigenetic inheritance.

Yet, while these studies provide critically important information on natural patterns of epigenetic variation, they do not provide evidence of the three ingredients required for adaptive evolution. For example, as we noted in chapter 5, in many such studies we don't have information on whether the epigenetic patterns are even transmitted across generations, or if these patterns are selectively important. As a result, a skeptic strongly committed to the genocentric view of evolution might rightly ask how such studies differ from those measuring any phenotypic trait like beak size or blood pressure. Such traits also surely differ among populations, and we might expect them to be more divergent than genotypes because they will be subject to environmental influences that differ among populations. In fact, a correlation between epigenetic patterns and environmental conditions could result either from such plasticity or from genetic adaptation if changes in the epigenome are a genome's way of building a phenotype appropriate for the environment. Of course, the skeptic's view is simply a different interpretation of the evidence, a very common situation in science. At the same time, however, it might be worth bearing in mind Marcello Truzzi's dictum that "extraordinary claims require extraordinary proof."

While these empirical hurdles will require a great deal of effort and ingenuity to overcome, it is also useful to explore the potential evolutionary implications of extended heredity theoretically. In the next chapter, we will ask whether our extended heredity framework could help in tackling some of the most challenging questions in evolutionary biology.

9

A New Perspective on Old Questions

> Questions are bigger than answers.
> —Stuart Firestein, *Ignorance*, 2012

Armed with the conceptual tools of extended heredity, in this chapter we will revisit a few of the most challenging and long-standing puzzles in evolutionary biology. We will not offer definitive answers to any of these questions. We merely hope to show that the extended heredity framework offers a fresh perspective, allowing us to see old questions in a new light and revealing potentially fruitful but hitherto unexplored directions for investigation. Extended heredity changes how we think about these problems because it alters some of the basic assumptions that have guided thinking over many years, such as the assumption that new heritable variation is generated exclusively by rare genetic mutation, and that environmental effects cannot be transmitted to descendants.

We have structured this chapter as a series of case studies, with topics arranged in a rough sequence from small-scale "microevolutionary" processes that occur over a few generations to large-scale "macroevolutionary" processes that span millions of generations and generate the broad diversity of living things. Other researchers have begun to apply similar ideas to a range of other evolutionary questions.[295] Although our discussion of these questions is speculative and preliminary, we hope that these examples will demonstrate the potential for extended heredity to enrich our understanding of evolution.

THE ENDLESS CHASE BETWEEN PARASITES AND HOSTS

Parasites make their living at the expense of the hosts that they infect. No doubt this fact is familiar to anyone who has suffered from an infectious disease like the flu or a simple cold. But this antagonism between host and parasite has important evolutionary implications as well. Any host type that can resist infection by a parasite will enjoy an increased reproductive success relative to other hosts, and so it will increase in abundance. Similarly, any parasite type that can evade such host defense mechanisms and cause infection will enjoy a greater reproductive success than other parasites. In this way, the host-parasite antagonism sets the stage for repeated cycles of host adaptation and parasite counter-adaptation. If this coevolutionary dynamic between species is relatively well matched, then neither party will gain the upper hand. Instead, a sort of perpetual chase will occur that evolutionary biologists have called a "Red Queen" dynamic. This terminology comes from Lewis Carroll's *Through the Looking-Glass* in which the Red Queen tells Alice that "it takes all the running you can do, to keep in the same place."[296]

The Red Queen provides an evocative metaphor, but what sorts of adaptations and counteradaptations do hosts and parasites actually evolve in real biological populations? The simplest examples were first discovered in plants and their pathogens by Harold Flor[297] in the 1940s and 1950s. Many parasites produce what are called *effector proteins* during an infection. Although the parasite can usually get by without these proteins, their production tends to enhance the ability of the parasite to replicate. In turn, plants have evolved resistance mechanisms that recognize these effector proteins. When the protein is recognized, the plant mounts a vigorous defensive response that prevents the infection from taking hold. Flor proposed that genes for effector proteins in parasites are often matched by genes for the recognition of these proteins in plants. This mechanism of adaptation and counteradaptation is therefore called the gene-for-gene mechanism.

The gene-for-gene mechanism has been the subject of considerable study over the past several decades.[298] Indeed, the basic logic underlying this form of host-parasite interaction has formed the basis for many programs of crop breeding aimed at producing plants resistant to pests. It is also the basis of an important evolutionary hypothesis for how genetic variation is maintained in populations. The Red Queen coevolutionary dynamic ensures that rare genotypes in both host and parasite are evolutionarily advantageous, and so natural selection drives the maintenance of genetic variation.

One interesting and important pathogen that has been the subject of extensive research is *Phytophthora sojae*. This pathogen infects soybean plants and can cause extensive crop losses through stem and root rot. The potential devastation that can be wrought by such a pathogen is made clear by the fact that *P. sojae* is closely related to *Phytophthora infestans*, a pathogen of potatoes that ran rampant in Ireland in the late 1840s, causing the Great Famine.[299]

In many ways, the *P. sojae*–soybean interaction seems like a poster child for the gene-for-gene mechanism. Alleles in *P. sojae* of a gene called *Avr3a* have been identified that code for different effector proteins.[300] For example, one allele called $Avr3a^{P6497}$ codes for a particular signal protein that is 111 amino acids long. In response, soybean plants can carry an allele called *Rps3a* that recognizes this effector protein and neutralizes the infection. Taking the Red Queen cycle full circle, a different allele of the *Avr3a* gene has been found in some *P. sojae* individuals, called $Avr3a^{P7064}$, that allows the parasite to evade the *Rps3a* defensive response. Thus, pathogens carrying the $Avr3a^{P7064}$ allele are able to cause disease, even in plant populations carrying the *Rps3a* defensive allele. As an illustration, Box 9.1 examines the rate at which this disease-causing allele is predicted to spread in a population of soybean plants that carry the *Rps3a* defensive allele.

Box 9.1. The Rate of Spread of a Disease-Causing Allele Such as $Avr3a^{P7064}$

To simplify our discussion let's use E to denote the $Avr3a^{P6497}$ allele because it produces the *effector* protein, and D to denote the $Avr3a^{P7064}$ allele because it does not and thus can evade the defensive response of the plant and cause *disease*. *P. sojae* is a diploid eukaryotic organism and so, like humans, each individual carries two copies of every gene in each somatic cell. Thus there are three possible parasite genotypes: EE, ED, and DD. Parasites with genotype EE will have essentially zero reproductive success because they produce the effector protein and thus are recognized and neutralized by the plant's *Rps3a* defensive allele. On the other hand, DD genotypes will evade the plant defense and so will cause disease, producing some number, say W, of offspring. Now we might expect ED genotypes to produce half as much effector protein as EE genotypes and so, for simplicity, let's suppose that they cause partial infections that result in $W/2$ offspring.

Now imagine a situation in which the E and D alleles are at equal frequency. If all genotypes are formed randomly, then the frequency of the three genotypes will be[301]

EE: 25%   ED: 50%   DD: 25%

EE individuals do not reproduce at all, and each ED individual produces half as many offspring as each DD individual. Therefore, overall the ED segment of the population will contribute an equal amount to the total reproduction of the population as the DD segment (because, even though each ED individual produces half as many offspring, they are twice as abundant as DD individuals: 50% vs 25%). The DD segment of the population is made up entirely of D alleles while the ED segment is made up of 50% of D alleles (and 50% E alleles). Therefore, after one generation the frequency of D will be

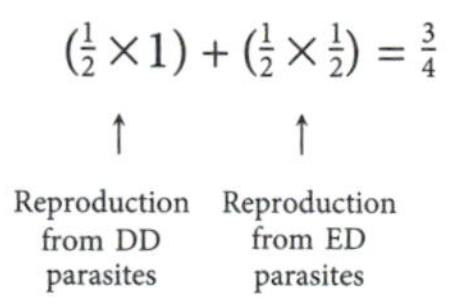

or 75%.

It is also instructive to use the theory developed in chapter 6 to address this question more generally.[302] The covariance term in the Price equation (see chapter 6) is calculated over the three genotypes giving $\text{cov} = (1 - p)/2$. And we also have $E[wd] = 0$ for the second term because

Continued on page 162

there is no change in allelic state from parent to offspring. Therefore, we get

$$\Delta p = \frac{1}{2}(1-p) + 0$$
$$\uparrow \qquad \uparrow$$
$$\mathrm{cov}(W,z) \quad E[wd]$$

or simply

$$\Delta p = \frac{1}{2}(1-p) \qquad (1)$$

For example, if we use a starting frequency of $p = 0.5$ as above, we obtain $\Delta p = 0.25$. Remembering that this represents the *change* in allele frequency, the new allele frequency is $p + \Delta p = 0.5 + 0.25 = 0.75$ as above.

Of course, things are never as simple as they first seem. Despite the *P. sojae*–soybean interaction being a beautiful example of the gene-for-gene mechanism, the *P. sojae* parasite is also now known to carry epialleles at the *Avr3a* gene.[303] For example, *P. sojae* parasites can evade the plant *Rps3a* defense simply by silencing the allele that they carry at the *Avr3a* locus through epigenetic mechanisms rather than swapping it out for a different allele. Furthermore, this epigenetic silencing is stably transmitted from parent to offspring. To make matters even more interesting, when a *P. sojae* individual inherits a normal epiallele (that is, an epiallele that does not silence the gene) from one parent and a silencing epiallele (that is, an epiallele that shuts down expression of the gene) from the other, the normal epiallele tends to become silenced as well before it is passed on to the next generation, appearing to violate the Mendelian rules of segregation (figure 9.1).[304]

The peculiar inheritance pattern of the *Avr3a* epialleles has very important implications. The epigenetic silencing allows the pathogen to evade a plant *Rps3a* defense and cause disease, and this silencing can also be spread to normal epialleles within an individual during reproduction. This means that in principle a disease-causing silencing epiallele could spread extremely quickly, even through a soybean population where all individuals carry the *Rps3a* defensive allele (figure 9.2). In the

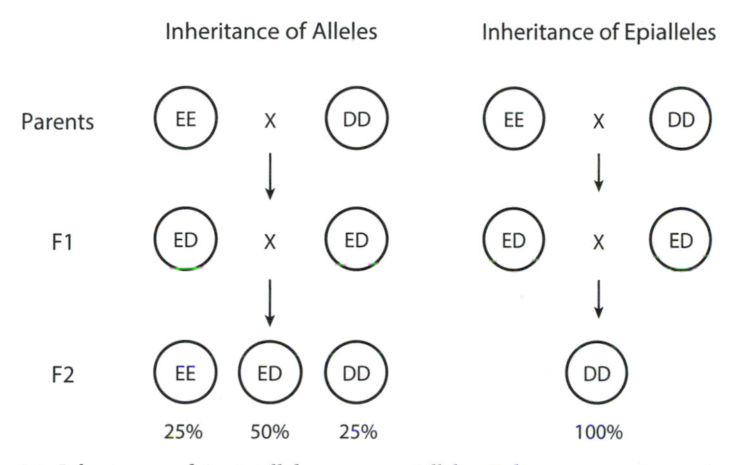

Figure 9.1. Inheritance of *Avr3a* alleles versus epialleles: E denotes an active variant that produces an effector protein while D denotes a variant that does not and so causes disease. *Left* (alleles): F1 offspring inherit an active and an inactive allele. Crosses between these F1 heterozygotes then produce the Mendelian ratio 25% : 50% : 25% in the F2 generation. *Right* (epialleles): F1 offspring inherit a normal and a silencing epiallele. The silencing epiallele then silences the normal epiallele before transmission, resulting in 100% of the F2 generation carrying only silencing epialleles.

most extreme case the frequency of the disease-causing silencing epiallele could reach 100 percent within a single generation (Box 9.2).

Box 9.2. The Rate of Spread of a Disease-Causing Epiallele

Let's use E to denote the normal epiallele that produces the effector protein, and D to denote the silencing epiallele that switches off production of this protein. Thus there are again three possible parasite types: EE, ED, and DD. As in Box 9.1, EE parasites have zero reproductive success, DD parasites produce *W* offspring, and ED parasites produce *W*/2 offspring.

Now if, as in Box 9.1, we imagine a situation in which the E and D epialleles are equally frequent, then the frequency of the three types is again

EE: 25% ED: 50% DD: 25%

All EE individuals do not reproduce, and each ED individual produces half as many offspring as each DD individual as before. Therefore,

Continued on page 164

as in Box 9.1, overall the ED segment of the population will contribute an equal amount to the total reproduction of the population as the DD segment. Now, however, when an ED individual reproduces, there is some probability that the normal epiallele E is silenced before being transmitted. Let's use κ for the probability of such silencing. The DD segment of the population is made up entirely of D epialleles as in Box 9.1, but the ED segment is now effectively made up of a fraction $\frac{1}{2}+\frac{1}{2}\times\kappa=\frac{1}{2}(1+\kappa)$ of D epialleles (and $\frac{1}{2}\times(1-\kappa)$ of E epialleles). Therefore, after one generation the frequency of D will be

$$\underbrace{(\tfrac{1}{2}\times 1)}_{\substack{\text{Reproduction}\\\text{from DD}\\\text{parasites}}} + \underbrace{(\tfrac{1}{2}\times\tfrac{1}{2}(1+\kappa))}_{\substack{\text{Reproduction}\\\text{from ED}\\\text{parasites}}} = \tfrac{3}{4}+\tfrac{1}{4}\kappa$$

which is somewhere between 75% and 100% depending on the probability of silencing, κ. Thus, in principle, the disease-causing silencing epiallele could reach a frequency of 100% in a single generation.

As in Box 9.1, we can use the theory developed in chapter 6 to address this question more generally.[305] The covariance term reflects differential reproductive success of the three types, and since this is the same regardless of whether the phenotype is determined genetically or epigenetically, we again have cov = $(1 - p)/2$. Now, however, because a silencing epiallele can silence the normal epiallele in heterozygotes, there is a difference in transmission fidelity between the two. As a result, the second term in the Price equation becomes $E[wd]=\frac{1}{2}\kappa(1-p)$. Therefore, we get

$$\Delta p = \underbrace{\frac{1}{2}(1-p)}_{\text{cov}(W,z)} + \underbrace{\frac{1}{2}\kappa(1-p)}_{E[wd]}$$

or more simply

$$\Delta p = \frac{1+\kappa}{2}(1-p) \qquad (2)$$

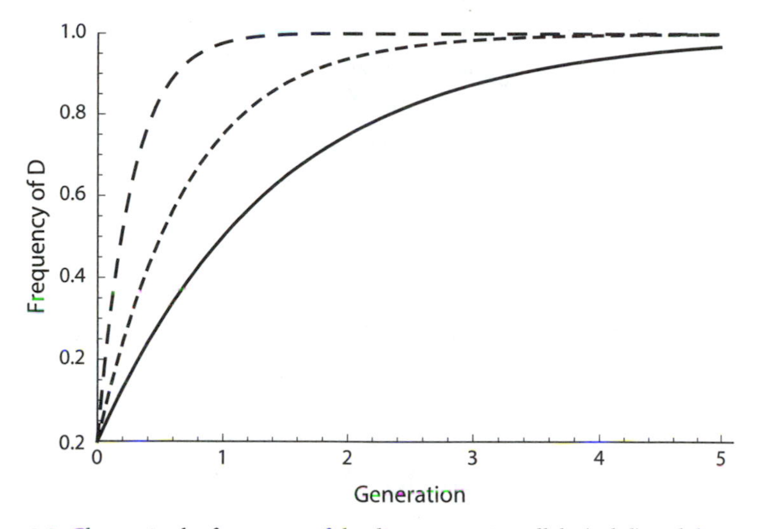

Figure 9.2. Change in the frequency of the disease-causing allele (solid) and disease-causing "silencing" epiallele (dashed) over five generations. Top dashed curve has $\kappa = 0.9$. Other dashed curve has $\kappa = 0.5$.

The predictions in Box 9.2 are speculative at this stage since the epiallelic variants of the *Avr3a* gene have only recently been described and much remains to be learned. However, the potential implications are significant. In addition to the possibility of extremely rapid spread of disease, the existence of such epialleles means that monitoring pathogen populations for their ability to cause disease solely by screening for genotypes will be inadequate. In doing so, we might never recognize that a disease-causing variant has appeared in the population until it is too late.

It is also interesting to contemplate whether the epigenetic silencing mechanism could be induced by environmental conditions. If so, then this would be extremely problematic in terms of breeding crops for resistance since an environmentally induced epigenetic silencing of pathogen *Avr3a* alleles might then instantly erase all the gains made through the breeding program. Instead, it might be interesting to explore epigenetic interventions designed to reactivate silenced alleles in the parasite so that they would then be susceptible to host defenses.

Finally, our brief considerations so far have focused solely on the parasite. Given the mounting evidence for nongenetic inheritance in complex eukaryotes, it is only natural to expect that plants like soybeans might also transmit nongenetic information to their offspring that affect their ability to resist infection.[306] The *Rps3a* allele is perhaps the best-described resistance mechanism, but epiallelic variants or other forms of nongenetic inheritance may also exist in soybeans, allowing the plant a means of rapid (and potentially non-Mendelian) defense. If so, the consequences for things like the Red Queen dynamic,[307] the practicalities of crop breeding, and the ability of host-parasite coevolution to drive the maintenance of genetic variation will all need to be examined with fresh eyes.

NATURE'S ENIGMATIC BEAUTY CONTEST

In the previous section, we considered coevolution between organisms of different species, but a great deal of interesting coevolution also occurs between the sexes within species. The classic puzzle of sexual coevolution is the evolution of female mate preferences and male displays—a question to which Darwin devoted much of his misleadingly titled second book, *The Descent of Man*,[308] and that continues to be a subject of great interest and a source of heated controversy to this day.

Fascination with mate choice stems from the observation that female animals are often remarkably selective about the males they accept as mates and the elaborate and grotesque displays that males of many species have evolved to woo females. The iconic example is the enormous, garish tail-feather display of the male peacock and the rapt attention that peacock females devote to it, observing and seemingly assessing its beauty like fickle cognoscenti at a fashion show. Other birds offer even more striking examples. Male birds of paradise in the rain forests of Papua New Guinea and Indonesia sport bizarre tail feather ornaments and splendid colors that they display to females through dazzling courtship dances. Male bowerbirds in the tropical forests of northern Australia build astonishingly complex structures from straw and adorn the surrounding area with colored objects to attract females. Less widely appreciated but equally spectacular are the displays of some male mammals, reptiles, fish, insects, and spiders. In fact, the tiny peacock spider, whose males unfurl brilliantly colored abdominal flaps as part of their

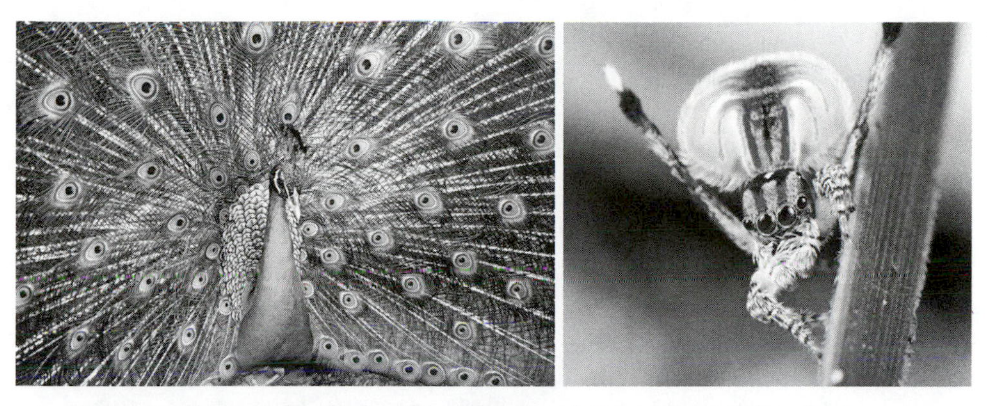

Figure 9.3. The courtship display of the male peacock *Pavo cristatus* (*left*) and male peacock spider *Maratus volans* (*right*). Could nongenetic inheritance help to explain why peahens and female spiders prefer to mate with the most attractive males? (*Left photo*: Jyshah Jysha; *right photo*: Jurgen Otto)

elaborate courtship, may be poised to displace the peacock as the iconic example of sexual display (figure 9.3).[309] (And, for anyone who needs reminding that beauty really is in the eye of the beholder, there is the hooded seal, *Cystophora cristata*, whose males display to females by inflating a sack of nasal skin into a huge pink balloon that protrudes from their left nostril.)

Observations of such species raise a basic question that has long troubled biologists: Why are females choosy about their mates? Darwin believed that females choose males that are "vigorous and well armed, and in other respects the most attractive" because mating with such males would enhance females' reproductive success.[310] But what do females actually *gain* by choosing? After all, it would be faster and simpler to mate with the first male that comes along. The answer to this question is straightforward in species where females acquire obvious goods or services from their mate—that is, where males care for offspring, provide females with a breeding territory, or hand over a "nuptial gift" of prey or nutrient-packed glandular secretions. Individual males vary in the quality of resources that they provide, so females in such species would do well to be discriminating. The problem is that, in most species, males do not appear to provide females or their offspring with any goods or services whatsoever. Males simply do the deed and move on, with the entire association between female and male sometimes lasting just seconds.

And yet, such species furnish some of the most spectacular examples of female choosiness and male display. Indeed, as though to deliberately taunt puzzled biologists, displaying males and choosy females of some of these species gather en masse at special show grounds called "leks," which tend to be barren patches utterly devoid of any food or shelter.

Generations of biologists have racked their brains trying to understand the evolution of female choice and male displays in such species, and the most popular hypothesis has been that choosy females are shopping for "good genes" for their offspring. The idea is deceptively simple. Sexual displays like the tail of the male peacock or the tail flap of the peacock spider should reveal the quality of the male's genes—after all, a sickly, low-quality male that carries harmful mutations will be in poor physical condition and will not be able to put on a truly spectacular show. So females that select the most attractive males will benefit because their offspring will get "good genes" from their father, and will thus inherit his health, vigor, and attractiveness. There's just one catch: for a "good genes" mechanism to work, populations must harbor plenty of genetic variation for fitness, and it's far from clear where all this variation would come from. "Bad genes" are constantly weeded out by natural selection, and genetic mutation appears to be too rare to maintain the needed level of genetic variation. Population-genetic models therefore show that genetic variation in fitness is quickly depleted, leaving only individuals with "good genes." But if just about every male carries "good genes," then females would seem to have very little to gain through mate choice—a conundrum that has been dubbed the "paradox of the lek."

Yet, there is another dimension to this story: whatever the level of genetic variation for fitness, there's no doubt that phenotypic variation is plentiful. Within every species, males vary enormously in the quality of their displays, and much of this variation undoubtedly reflects the environment in which they developed—especially the quality and abundance of dietary resources and exposure to various sources of stress. Could such variation hold the key to the mystery?

From the Modern Synthesis perspective, all this phenotypic variation counts for nothing unless it reflects genetic variation—that is, unless attractive males are attractive because they carry "good genes." This is because it's assumed that the environmental component of phenotypic variation is not transmitted to descendants. From this perspective, it matters not a whit what kind of food your father ate or how stressed out he was; only the genes that you inherit from your father can influence your features and fitness.[311] But the picture changes under extended heredity because, as we have already seen, nongenetic inheritance allows for the transmission of environmental effects to offspring. This suggests a potential solution to the paradox of the lek: perhaps females are choosy because, by mating with attractive males, they can ensure that their offspring will obtain environmentally induced nongenetic benefits? However, three conditions must be satisfied for this to work. First, it must be possible for environmental variation in male quality to affect offspring fitness through nongenetic paternal effects. Second, unlike genetic variation in fitness, nongenetic variation in male quality must be maintained despite persistent directional selection. Third, nongenetic paternal effects must be able to substitute for paternal genes in promoting the evolution of female preferences.

The first condition is clearly satisfied: as we saw in chapters 4 and 5, effects of paternal environment on offspring have been reported in many animals. For example, in neriid flies, a male's nutrition as a maggot influences the body size of the offspring that he sires as an adult via factors transferred in the seminal fluid, and larger offspring are likely to enjoy a fitness advantage. The second condition is also easily satisfied because much phenotypic variation is generated by the environment. There are good and bad food patches, and many unlucky individuals invariably end up in the latter. There are many sources of stress as well. Environmental heterogeneity will therefore guarantee an abundant and never-ending supply of sickly, low-quality individuals with unattractive displays.

But the third condition is less straightforward. Many nongenetic paternal effects appear to fade out after just one or two generations, so it's not obvious that such effects could drive the evolution of female preferences in the same way as stable "good genes." Therefore, to determine whether the third condition could be satisfied, we created a virtual world (that is, a mathematical model) in which males are randomly distributed among good and bad food patches, and their luck in this raffle then determines their condition (that is, their health and vigor). We also imagined that, as in neriid flies, males in high condition transmit a beneficial nongenetic paternal effect that gives their offspring an advantage. To see how this would influence the evolution of female preference, we gave our virtual animals a genome containing a mate choice locus with two possible alleles—an "indiscriminate" allele causing females to accept any male as a mate, and a "preference" allele causing females to prefer males in high condition. If nongenetic paternal effects can select for female preference just like "good genes," then the "preference" allele should increase in frequency and the "indiscriminate" allele should decrease in frequency.

The results were clear.[312] Because females mating with high-condition males produced more fit offspring and therefore enjoyed higher fitness themselves, the nongenetic paternal effect strongly favored the "preference" allele and drove the evolution of female mate choice in our virtual population. In real animals, choosy females typically use condition-dependent ornaments and signals (such as the courtship displays of male peacocks and peacock spiders) to identify high-condition males, so a preference allele like the one that spread through our virtual population would tend to drive the evolution of such condition-dependent male traits as well. High-condition males produce the most spectacular displays and are therefore most attractive to females. These male traits would likely have a genetic basis—a genetic mechanism that causes the healthiest, most well-fed males to invest extra resources in their sexual displays. Yet, our model shows that the evolution of these genetic traits (along with female preference itself) could be driven by nongenetic inheritance of paternal condition. Indeed, this effect was possible because we assumed that heritable variation in fitness was continually regenerated by the environment, ensuring that every generation of females encountered males that would produce high- and low-fitness offspring, and giving choosy females that preferred high-condition males an advantage over females that mated indiscriminately. This shows that a common type of nongenetic paternal effect—call it the "inheritance of acquired condition," or more simply "condition transfer"—has the potential to resolve the long-standing paradox of the lek. Such paternal effects have been reported in a wide variety of animals, providing a new, general hypothesis for the evolution of female preferences and male displays.

To better appreciate why this nongenetic paternal effect avoids the paradox of the lek, it's useful to compare this outcome with some alternative scenarios. First, what happens if condition is determined by a genetic locus, as in the classic "good genes" case? Here, females mating with high-condition males (which carry high-fitness alleles at the condition locus) also benefit by producing fitter offspring. However, persistent selection quickly drives the high-fitness allele to fixation. Although new low-fitness alleles occasionally appear in the virtual population through mutation, genetic mutation is simply too rare to maintain selection for preference—that is, mutant males are so infrequently encountered that females stand to gain very little by discriminating against them. Because we assumed that choosing mates was costly for females (as it must often be for real animals), we found that the preference allele was rapidly lost in this situation. This result illustrates the classic lek paradox.

It's also interesting to examine what happens if male condition is determined by an epiallele that is not sensitive to the environment but is subject to a much higher rate of spontaneous mutation than is typically assumed for genetic loci. Can a higher mutation rate that supplies more heritable variation in fitness suffice to maintain costly female preference? The results in this case were intermediate between the two other scenarios: we found that costly preference was maintained, but only when the epimutation rate matched the strength of female preference. This is because in the epigenetic case, just as in the "good genes" scenario, selection could deplete heritable variation, so preference could only be maintained in the long run if the epimutation rate was sufficiently high.[313] That is, the preference allele was lost unless epimutation continually supplied enough low-quality males to compensate for the costs of choosing mates.

To us, these findings are interesting because they show that even a very unstable nongenetic factor, such as an environmentally induced paternal effect that fades out after just one or two generations, can still influence evolution, contradicting the intuition that only stable hereditary factors can play an evolutionary role. Indeed, we found that the very transient paternal environment effect selected for and maintained female preferences more readily than semistable epialleles, and far better than very stable genetic alleles.

Because we assumed that female preference is determined genetically (as it is in many insects and other animals),[314] this nongenetic solution to the lek paradox also provides an example of a feedback loop between genetic and nongenetic inheritance systems. As we saw in chapter 7, a special form of such feedback—called "gene-culture coevolution"—is thought to have played an important role in human evolution. The potential role of nongenetic paternal effects in the evolution of female mate choice shows that similar feedbacks could occur in other species and outside the cultural context.

But perhaps the story doesn't end there. As we mentioned in chapter 5, neriid fly males can transmit their environmentally induced condition not only to their own offspring but also, via factors in the seminal fluid, to offspring of other males that mate two weeks later with the same female. The potential for such "telegony" effects means that females could benefit from mating and mate choice even when they have no eggs ready to be fertilized, because beneficial nongenetic factors received from high-condition males could enhance the fitness of offspring produced later on. In species where telegony occurs, females could therefore evolve mate preferences that change over the course of the reproductive cycle. When outside their fertile phase, females could choose males that confer advantageous nongenetic seminal fluid-borne factors, such as seminal RNAs or proteins that can affect the development of immature ovules; conversely, when fertile, females could choose males that provide advantageous sperm-borne factors, such as genes or sperm-borne epialleles. Such cyclical preferences could be predicted to evolve when seminal fluid-borne and sperm-borne effects are imperfectly correlated, and when females can detect and assess distinct male signals of each type of benefit. Such effects could perhaps even select for choosiness in males, since males might benefit by detecting chemical cues from a female's previous mates and rejecting females that had already mated with low-quality males whose seminal fluid could harm the female's future offspring. Thus, if future research confirms that telegony plays a role in natural populations and occurs in other species, this phenomenon could have interesting implications for the evolution of female and male sexual behavior.[315]

WHO NEEDS SEX?

While the evolution of mate choice has long puzzled biologists, an even deeper question is why so many organisms bother to reproduce sexually at all. This may seem like an odd question, but consider this:

some organisms get along perfectly well without ever having sex, or with just occasional bouts of sexual reproduction interspersed between generations of asexual reproduction, employing a variety of physiological mechanisms that enable unfertilized eggs to develop into viable offspring that are clones or near-clones of their mother. What's more, sexual reproduction comes with a long list of drawbacks and costs. First and foremost among these is the fact that asexual organisms produce only daughters, and because each daughter can produce offspring of her own, an asexual population can grow twice as fast as a sexual one.[316] Then there's the problem of having to find a mate and, for females in many species, the converse problem of fighting off unwanted suitors and avoiding the harm of excessive mating. Yet, despite these costs, sexual reproduction is widespread in complex organisms and the only mode of reproduction in most animals.

Most attempts to resolve this mystery have focused on a particular feature of sexual reproduction: the generation of novel genotypes. Sexual reproduction occurs in a mind-boggling variety of ways, but a feature shared by all of them is genetic recombination followed by the mixing in the offspring of genes from two parents. Recombination occurs during the formation of gametes, when the homologous chromosomes pair up, link together at several points, and then exchange corresponding chunks to form new chromosomes. These new chromosomes contain new combinations of alleles, drawn randomly from the original chromosomes like cards from a deck. At fertilization, recombined chromosomes from the egg and sperm come together to form a new, unique genome in each offspring.

This tendency to generate new gene combinations is widely seen as a potential benefit of sex. By shuffling genes into new combinations, sex can create both advantageous genotypes that will enjoy high fitness and disadvantageous genotypes that will be efficiently weeded out by natural selection. Sex therefore has the potential to promote adaptation as well as contribute to the purging of deleterious mutations from populations. As an example, imagine that two different individuals in an insect population carry potentially beneficial mutations, with one individual having a greenish color and the other individual having a leaf-like shape. In an asexual population, these alleles have no way of coming together in the same genome unless both mutations happen to occur in the same individual or lineage, but in a sexual population mating and recombination might produce a genome that contains both alleles and gives its bearer much-enhanced camouflage. Likewise, asexual lineages will tend to accumulate mildly deleterious mutations, each of which is not severe enough to lead to death or sterility but whose combined effects will gradually erode fitness over many generations. In a sexual population, mating will generate unlucky individuals with a disproportionate number of deleterious mutations in their genome, and these individuals will be selected out of the gene pool. It will also generate lucky individuals that are relatively free of mutations, and it's these individuals that will thrive and breed.

Yet, these hypothesized benefits of sexual reproduction do not seem entirely adequate to account for the prevalence of sex in nature. The problem is that all such explanations are based on restrictive assumptions about the genetics and ecology of populations, such as a high mutation rate, importance of interactions between different alleles within the genome, or rapid environmental change. While such explanations can account for the prevalence of sexual reproduction over asexuality in some circumstances, their explanatory breadth seems limited. Could extended heredity help to resolve the mystery of sexual reproduction?

Conventional thinking about the evolution of sex has focused on the role of genes, but of course sexual reproduction also universally involves the transfer of nongenetic factors—the cytoplasm and epigenome of the egg and sperm (figure 9.4). The transfer of these nongenetic components is an obvious and well-known feature of fertilization, but its implications for the evolution of sex have not been thoroughly explored.

Consider the interesting models, developed by Lilach Hadany and coworkers, showing that sexual reproduction is especially advantageous and likely to evolve if it occurs in a condition-dependent manner, such that sickly females reproduce sexually more often than healthy females do.[317] This makes intuitive sense. If high-condition females have good alleles and well-functioning allele combinations in their genomes, they will derive little benefit from mixing their alleles with alleles from other individuals. Conversely, if low-condition females carry deleterious mutations or poorly functioning combinations of alleles, their best strategy will be to mix genetically with other individuals because by doing so they can produce offspring with fewer bad genes or poorly functioning

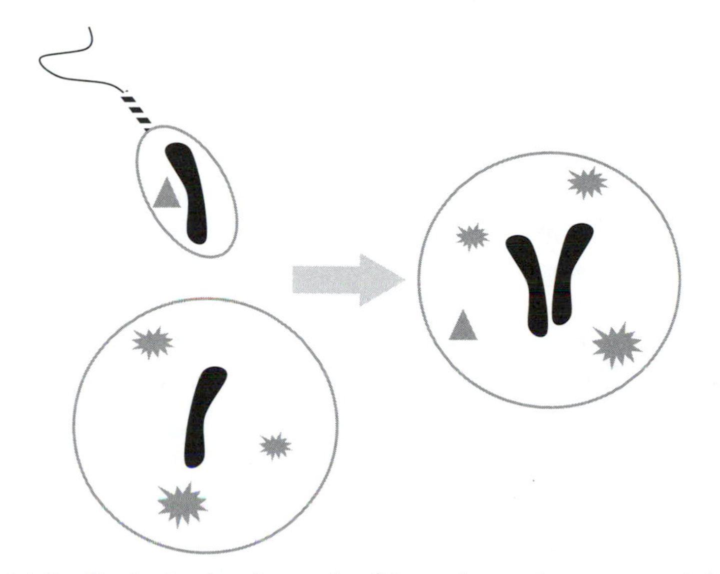

Figure 9.4. Fertilization involves the transfer of egg- and sperm-borne nongenetic factors to offspring alongside genes.

gene combinations. While Hadany's analysis helps to explain why sex might be advantageous, it also leads to a paradoxical conclusion: the best strategy for a high-condition female is to simply clone herself. The most plausible outcome predicted by her models is a plastic strategy where individuals assess their own genetic quality and reproduce sexually only if the load of deleterious mutations in their genome exceeds some threshold. We should therefore expect to find many species in which low-quality individuals always reproduce sexually while high-quality individuals always reproduce asexually. But such species are extremely rare.

Incorporating the nongenetic dimension of sexual reproduction into Hadany's theory may help to resolve this conundrum. It's clear that a low-condition individual can benefit by combining not only its genome but also its cytoplasm and epigenome with that of another individual so as to replace both genetic and nongenetic components of its poorly functioning cellular machinery in its offspring. For example, a low-condition female might benefit by supplementing her eggs with high-quality RNA, proteins, epialleles, or centrioles from a male. The nongenetic dimension of fertilization could therefore augment the advantages of genetic recombination and exchange, providing an additional benefit of sex. More importantly, the nongenetic dimension may extend the benefits of sexual reproduction to all genotypes in the population. Because many nongenetic factors are subject to environmental induction, no genotype is immune to the effects of a poor-quality environment, and all individuals, not just those that carry deleterious mutations, can experience poor condition. Moreover, aging results in bodily deterioration in all individuals that survive long enough. Thus, given that nongenetic factors vary in quality independently of genes and are prone to deterioration over every individual's lifetime, the ability to reproduce sexually is likely to be advantageous for all genotypes.

Extended heredity also raises some new questions. Hadany's models and other theory based on the benefits of genetic mixing implicitly assume that females and males are functionally equivalent: the models simply posit mating individuals that exchange genetic material. This assumption is reasonable from the genetic perspective because the zygote receives haploid genomes containing mostly equivalent chromosomes from its mother and father. But this assumption doesn't hold when considering the nongenetic dimension of fertilization, because male gametes almost always contain less cytoplasm than female gametes do.

This difference in the nongenetic content of eggs and sperm points to an interesting prediction—namely, that an individual's condition can determine not only whether to reproduce sexually but also what sex role to play. We mammals are used to thinking of an individual's status as male or female as a feature fixed at fertilization, but many animals (like turtles and crocodiles) have the ability to develop into either sex, and some (like certain fish) even have the ability to switch from one sex to the other. In such organisms, sex is not determined by the sex chromosomes; instead, sexual development typically occurs as a plastic response to some environmental, social, or physiological cue. Many other species are hermaphroditic, with every individual capable of playing either the male or female role. In such species, it's been suggested that an individual's condition can determine the most advantageous sex role to play because of the unequal intensity of sexual competition faced by females and males. When competition for mates is fierce, a low-condition individual may do better as a female than as a male because males generally have to compete more intensely to secure matings than females do. A sickly female can expect to produce at least some offspring, whereas a sickly male will probably fail to sire any offspring. But extended heredity suggests an additional dimension to this problem that can reverse this prediction: because males contribute less cytoplasm to the zygote than females do, a low-condition individual might benefit by playing the male role so as to minimize its nongenetic contribution to its offspring. Given these opposing predictions, the outcome might depend on the relative importance of sexual competition and nongenetic inheritance as determinants of fitness.

Of course, eggs and sperm differ not just in the quantity of cytoplasm but also in the specific types of nongenetic information that they can transmit to the embryo. For example, as we saw in chapter 4, epigenetic factors in eggs and sperm do not appear to be equally likely to persist in the zygote. Likewise, egg and sperm contribute differently to the cytoskeletal structure of the zygote: the egg surely carries more structural information, but sperm can contribute potentially important elements such as centrioles. Whether an individual would do best by playing the female or male role in reproduction could therefore depend on the quality of many nongenetic factors. And if particular factors contributed by males are especially important for offspring fitness, then a high-quality individual may do best to play the male role so as to transmit these advantageous nongenetic factors to its offspring.

Extended heredity could have especially interesting implications for the origin and early evolution of sexual reproduction in primitive eukaryotes. Nongenetic inheritance may be particularly important in such organisms because their lack of a specialized germ line facilitates the transmission of structural, cytoplasmic, and epigenetic factors to offspring, and some authors have already considered the role of some of these nongenetic components, such as mitochondria,[318] in the evolution of sexual reproduction. Indeed, a classic model posits a central role for the cytoplasm in the origin of male and female reproductive strategies. Geoff Parker and colleagues showed that a scenario in which equal-sized cells fuse together and then divide into daughter cells is unstable because selection will favor cheaters that produce smaller cells and contribute a lesser share of cytoplasmic resources.[319] The cheaters that evolve in Parker's model are, of course, primordial males. While Parker and colleagues focused on the quantity of cytoplasmic resources, the quality of various nongenetic factors could play a role as well. For example, selection on sickly or senescent individuals to dilute their damaged cytoplasm with that of healthier individuals could have favored a protosexual strategy of fusion with other cells, and the benefits to such individuals of playing the "male" role so as to shed a greater proportion of their damaged cytoplasm could have contributed to the evolution of sexual differentiation. In such a scenario, the male strategy would benefit not only from the ability to produce more gametes but also through the production of higher-quality offspring.

WHY DO WE AGE?

Like sexual reproduction, aging (senescence) is something that most people take for granted. After all, inanimate objects deteriorate and break down over time, so on the face of it, there's nothing surprising in the fact that living organisms fall apart as well. The problem with this view is that species vary dramatically in aging rate, and this shows that aging cannot be a simple, inevitable outcome of mechanical wear and tear. Why, for example, do dogs age so much faster than their human owners, despite living in very similar conditions? Why do rats rarely see their second birthday, while many small bats and birds can live to see their tenth? How can we explain the fact that a tortoise named Harriet, collected by Charles Darwin on his visit to the Galápagos Islands in 1835, managed to outlive her famous collector by more than a century?[320] Such biological variation strongly suggests that aging is an evolved trait. Indeed, living organisms continually repair and regenerate their tissues, and this capacity would seem to offer an evolvable physiological mechanism for the regulation of aging. Perhaps Harriet outlived Charles because tortoises maintain and repair their tissues better than humans do.

To explain the evolution of aging, twentieth-century biologists invoked the implications of "extrinsic" sources of mortality—that is, accidents, predators, and other risks that are at least partly unavoidable. Extrinsic mortality ensures that organisms would have a finite life expectancy even if they did not age at all, and that individuals' health and performance at older ages has less importance for fitness than their health and performance at younger ages. Because of extrinsic mortality, traits expressed in old age are under weak selection simply because few individuals live long enough to express them, so that their average effects across all individuals are negligible. It is this inevitable discounting of old age in the cold calculus of fitness that is thought to allow aging to evolve, and its evolution is thought to involve several interrelated processes. To begin with, mutations with deleterious effects that only become apparent in old age will be largely invisible to natural selection because few individuals will live long enough to suffer their effects, and such mutations will therefore accumulate in the genome. Moreover, mutations that are advantageous in early life (for example, because they increase reproductive output) will be favored by selection even if they promote deterioration later on. For similar reasons, selection will also tend to limit the allocation of resources to the maintenance and repair of the body (soma)—that is, it will act against the pursuit of immortality.[321]

A thought experiment helps to grasp this idea.[322] Imagine an organism that allocates enough energy and resources to the repair of its bodily tissues to completely prevent aging, thereby making itself potentially immortal. Such an organism could perhaps survive for centuries in a zoo. However, its life expectancy would be much lower in the wild because it would eventually succumb to a predator or accident, and this would mean that some of the resources that it had invested in somatic repair were simply wasted. Individuals that invested less in repair and more in reproduction would tend to produce more offspring, and reduced investment in repair would therefore evolve. Ultimately, natural selection is expected to adjust investment in somatic repair to the level that will keep organisms healthy and vigorous for only as long as they can expect, on average, to avoid being killed by predators or accidents—that is, to an age corresponding roughly to the life expectancy determined by the extrinsic mortality rate.

This classic theory has considerable power. It explains why organisms deteriorate with age. It also predicts that higher predation risk should be associated with faster aging, potentially explaining why large-bodied species typically exhibit slower aging than small-bodied ones, why animals that possess effective defenses such as flight or shells tend to age more slowly than animals lacking such defenses, and why humans (with their technological defenses and safeguards) age more slowly than other mammals of similar body size. Yet, this theory has not provided a satisfactory explanation for some important aspects of variation in aging, such as striking differences in aging between populations seemingly subject to similar extrinsic mortality rates, and the considerable variation in the rate and pattern of aging observed among individuals within populations. This suggests that something might be missing from the classic theory, and we believe that clues to this missing dimension might be found in the advances made in recent years in the cell biology and epigenetics of aging. While these discoveries have informed medical research on aging, they have had almost no impact on the evolutionary theory of aging. Could extended heredity link these proximate and ultimate levels of analysis?

Several clues suggest such links. The first clue comes from recent studies led by Steven Horvath that show that the human epigenome undergoes striking changes with age, with certain genomic regions tending to become more heavily methylated and other regions tending to undergo demethylation. These changes are so consistent across bodily tissues and between individual people that they have been dubbed the "epigenetic clock." Consistent patterns of age-related epigenetic change occur in mice and rhesus monkeys as well.[323] In other words, while the genomic DNA sequence remains essentially unchanged throughout life, the epigenome undergoes profound modifications that parallel and probably causally contribute to the physiological deterioration that we associate with aging.[324] Of course, these changes in DNA methylation occur in parallel with changes in other epigenetic, cytoplasmic, and somatic factors.

The second clue is that, like aging itself, the epigenetic clock is highly sensitive to stress. For example, the human epigenetic clock is accelerated by obesity, severe psychological trauma, and urban poverty,[325] while mouse and monkey epigenetic clocks are slowed when the animals are fed a low-fat diet or given the drug rapamycin—two interventions that are well known to retard aging.[326] Such findings are consistent with evidence that identical twins become increasingly dissimilar epigenetically as they age;[327] if twins are exposed to different levels of stress during their lives, this may cause their epigenetic clocks to run at different rates. These studies are supported by evidence from many other species showing that stress promotes aging and implicating epigenetic dysregulation as a key molecular mechanism underlying these effects.[328]

The third clue is that, in organisms ranging from yeast to mammals, offspring quality tends to decline with parental age at breeding, with offspring of older parents tending to be sickly and short-lived.[329] Although such "parental age effects" could result from the accumulation of genetic mutations in the germ line, the consistency of the patterns and their sensitivity to environment suggest that nongenetic inheritance plays an important role. For example, in rotifers, offspring of old mothers fared less poorly if their mothers were fed a low-calorie diet.[330] The role of nongenetic inheritance in parental age effects is also consistent with the epigenetic clock. As we saw in chapter 4, some environmentally induced epigenetic changes are known to be heritable, and this means that the age- and stress-related epigenetic changes represented by the epigenetic clock could be a basis for parental age effects. If epigenetic changes occur in the germ line just as they do in other bodily tissues, and if older parents transmit part of their dysregulated epigenome to their offspring, those offspring could be expected to suffer ill health and reduced longevity. In essence, offspring of older parents might be born with a prematurely senescent epigenome. However, parental age effects need not be limited to epigenetic changes, nor even to factors transmitted through the germ line; any aspect of parental investment in offspring (for example, the cytoplasm of the egg and sperm, the intrauterine environment, milk quality or quantity, or parental behavior toward the offspring) that deteriorates as the parent senesces could cause offspring performance to decline with parental age. There is also evidence that parental age effects can accumulate over multiple generations; in *Drosophila melanogaster*, longevity is affected by both the mother's and maternal grandmother's age at breeding while, in neriid flies, longevity is also affected by the father's and paternal grandfather's age at breeding.[331]

Taken together, these clues point to a potential role for nongenetic inheritance in the evolution of aging. Consider what might happen if a population is subjected to an increased level of stress. This stress will accelerate the rate of epigenetic (and other nongenetic) dysregulation with age. Those individuals who manage to escape such stress-induced dysregulation will continue to enjoy the highest reproductive success, but despite this advantage the heightened rate at which dysregulation occurs means that a greater fraction of the population will nevertheless come to be dysregulated under stressful conditions. Now, if some of these altered epialleles are then transmitted to offspring, the stress may thereby also cause offspring quality to decline more rapidly with parental age. If such effects accumulate over multiple generations, then we would expect to see the affected population undergo phenotypic changes that resemble the evolution of accelerated aging: reproductive performance will decline more rapidly with age, and these changes will build up over generations. Indeed, the evolution of accelerated aging *will* have occurred, but it would have been underlain by environmentally sensitive epialleles rather than genes. And because these changes in aging would be heritable, they would be likely to persist for at least one or two generations even under stress-free conditions.

However, if elevated stress levels persist over many generations, these epigenetic changes could be expected to generate selection on genetic alleles and thereby kick-start the genetic evolution of aging as well. As parental age effects become more severe, selection will favor alleles that elevate reproductive effort in early life because individuals that breed at younger ages will benefit by producing offspring of higher quality. Moreover, just like elevated extrinsic mortality risk, increased stress will further reduce the strength of natural selection on older individuals—this time not because fewer individuals manage to survive to old age but because older individuals produce fewer viable, fertile offspring. The more rapid reduction in selection strength with advancing age will also promote alleles that down-regulate investment in somatic maintenance and, at the same time, will allow deleterious mutations with a late age of onset to accumulate. The affected population is therefore likely to undergo both epigenetic and genetic changes, and these changes will be mutually reinforcing, potentially resulting in a positive feedback loop that drives the evolution of aging. Note that this hypothetical process combines two key elements of the predicted evolutionary role of nongenetic inheritance: the propensity for nongenetic change to precede genetic change, and the tendency for nongenetic and genetic factors to interact in influencing the dynamics and direction of both phenotypic and genetic change.

This epigenetic hypothesis differs in important ways from the classic theory. First, the epigenetic model predicts that the evolution of aging can begin without change in the extrinsic mortality rate. All that's needed is a change in the level of stress that individuals experience; even nonlethal stress will accelerate the epigenetic clock, parents will transmit their epigenetic dysregulation to their offspring, and such effects will accumulate over multiple generations as individuals are born with an increasing load of deleterious epialleles. Second, since this process is initially mediated largely or entirely by epigenetic changes, its early stages should be reversible over a small number of generations if stress returns to its former level.

The epigenetic model therefore yields some surprising predictions. For example, stresses that are not necessarily associated with increased mortality (such as psychological trauma) might nonetheless have dire effects that accumulate over generations and resemble the evolution of accelerated aging. In populations subjected to such stresses, we may see not only direct effects on stressed individuals themselves but also on their descendants, and the effects of such stresses might persist for several generations even after stress levels decline. Similar processes could also occur in other animals. For example, snowshoe hares subjected to the psychological trauma of simulated predator attack produced offspring with reduced birth weight—an effect that could be mediated by stress-induced epigenetic changes.[332] On the other hand, the epigenetic model also predicts that some populations currently suffering from rapid aging could recover relatively quickly if stress levels are reduced; to the extent that the legacy of ancestral stress is epigenetic rather than genetic, the alleviation of stress could result in improved health over a much smaller number of generations than would be required for genetic evolution.

The epigenetic model could also help us understand within-population variation in aging rate. Even if all individuals carry similar genetic alleles, some individuals will invariably experience psychological or physiological stress and may transmit the resulting epigenetic dysregulation to their offspring. The legacy of ancestral stress could then manifest as variation in aging. In other words, the epigenetic model predicts that the evolution of aging could be a far more rapid and labile process than envisioned by the classic theory.

THE ORIGIN OF SPECIES

Despite the title of Charles Darwin's most famous work, *On the Origin of Species*, his landmark publication did not devote much attention to the processes by which new species arise. In retrospect, perhaps this is not surprising as the issue turns out to be considerably subtler than it first appears. For example, even to begin discussing the evolution of a new species requires that we first provide an unambiguous definition of what a species is. This alone is a contentious can of worms that has occupied the minds of evolutionary biologists for decades.[333]

For our purposes here, we will sidestep these philosophical land mines and adhere to a version of the so-called biological species concept. We will define a species as a group of individuals that has the potential to interbreed and that is reproductively isolated from other such groups. The evolution of a new species is then the process by which a group of interbreeding individuals splits off from an existing species and comes to be reproductively isolated from this ancestral group.

There are several ways in which new species are thought to evolve, but one relatively simple scenario harkens back to Sewall Wright's concept of an adaptive landscape, introduced in chapter 7. As we have seen, eventually a population is expected to approach a peak on the adaptive landscape because near a peak most individuals in the population have traits that confer high reproductive success. If the landscape contains many different peaks, however, then the particular peak to which a population initially ascends will be determined, in part, by where this population starts.

Now one way to envisage the formation of a new species is to imagine that part of a population on one peak—say peak A in figure 9.5—somehow manages to descend through the foothills, to traverse the valley of the adaptive landscape, and to then colonize a new peak—say peak B in figure 9.5. At this point, individuals from peak B will be nearly reproductively isolated from those of peak A because any offspring produced through an interpopulation mating will fall into the valley and thus have low fitness. From here it is then easy to imagine that further reproductive incompatibilities between the two populations might arise, eventually leading to complete reproductive isolation. Thus, a so-called peak shift from one peak to another can be viewed as one process that initiates the evolution of a new species.

But how can such a peak shift occur? After all, somehow a population must first pass through a low-fitness valley, and the very process of evolution by natural selection ought to prevent this from occurring. This problem has vexed evolutionary biologists for many years, and Sewall Wright himself was the first to appreciate the difficulty. His solution was

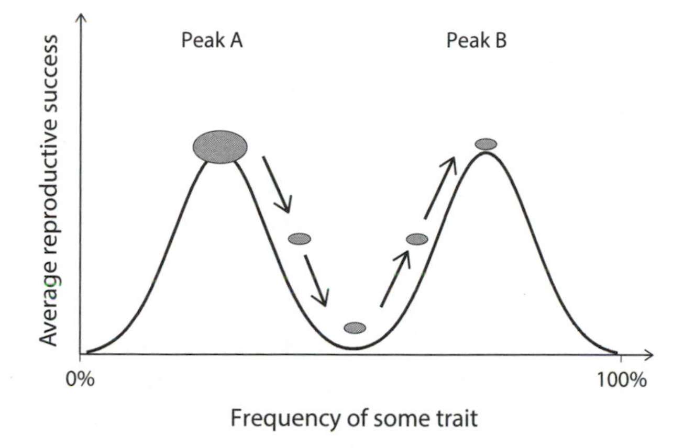

Figure 9.5. A schematic illustration of a peak shift. Part of the gray population located at Peak A traverses the fitness valley and then evolves toward Peak B.

to suppose that, if a population's size got small enough, then a peak shift might occur essentially by chance, even though natural selection would initially tend to oppose such a shift.

It is interesting to note that nearly twenty years ago, long before the study of nongenetic inheritance started to become fashionable, some researchers were already exploring how epigenetic inheritance might provide an alternative solution.[334] A more recent theoretical study by Filippos Klironomos and colleagues[335] has now more clearly driven home this possibility by showing how an interaction between genetic and epigenetic inheritance readily allows a population to move from one peak to another.

Although their simulation study is quite complex, the results are relatively straightforward to understand. They conducted simulations in which a population evolves on an adaptive landscape containing multiple peaks. In one set of simulations the population could evolve only via genetic change. In another set of simulations, the population could evolve via genetic change, epigenetic change, or both. The main difference between the genetic and epigenetic inheritance systems was that the latter had a larger mutation rate than the former.

In the simulations involving only genetic inheritance, populations tended to evolve to a local peak on the landscape and then remain there indefinitely, just as Sewall Wright had worried they would. In simulations that also allowed for epigenetic inheritance, populations again tended to evolve to a local peak on the landscape. After a relatively short period of time at this peak, however, populations often then shifted and evolved to a different, higher peak on the landscape.

What is it about the presence of epigenetic inheritance that facilitates such peak shifts? Klironomos and colleagues found that during the initial stages of adaptation the population evolves to a local peak primarily via epigenetic changes. This is because the epigenetic inheritance system has a greater mutation rate and this allows the population to explore different locations on the adaptive landscape more quickly than the genetic inheritance system. And for the same reason, even once a population has reached the summit of a local peak, the high mutational input of the epigenetic system continues to provide a means for the population to sample different locations on the landscape. Of course, most of these epigenetic mutants will be less fit than their parents and so will not leave many descendants, but the occasionally lucky mutant will arise that allows for a peak shift. This does not tend to happen in populations with genetic inheritance alone because the mutation rate is too low.

Although epigenetic variability allows for peak shifts, this is not the end of the story. The high epigenetic mutation rate of a population located at a peak allows it to effectively sample other locations on the landscape, but it also means that all individuals in the population will tend to have a compromised reproductive output because a substantial fraction of their offspring will carry mutations that drop them into a fitness valley. However, during the time that a population is at a peak through epigenetic adaptation, genetic changes can also continue to occur. Eventually the population hits upon a genotype that also codes for adaptation to this peak. Once such a genotype has appeared, individuals carrying this genotype will leave a greater number of successful offspring than individuals that are adapted through epigenetic means because they will endure a lower mutation rate. Consequently, the genetic inheritance system eventually displaces the epigenetic inheritance system as the substrate for adaptation.

In this way, the epigenetic inheritance system provides a means for a population to explore different locations on the adaptive landscape and so facilitates peak shifts, while the genetic inheritance system eventually takes over, allowing a population to solidify its location at a peak. Naturally, these theoretical findings are somewhat preliminary since they

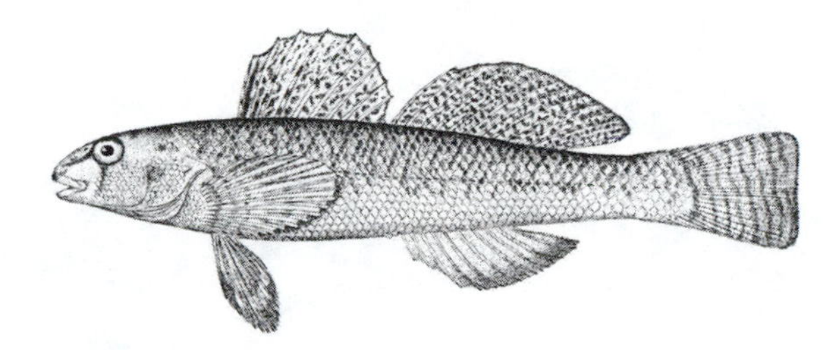

Figure 9.6. Illustration of a tessellated darter, *Etheostoma olmstedi*. (Source: Wikimedia Commons)

involve a specific set of assumptions about how the genetic and epigenetic systems interact. But such results have already started to motivate empirical research. Two studies in particular have begun to examine whether we see a pattern of epigenetic evolution followed by genetic evolution during the process of speciation.

One study involves a genus of freshwater fish species called darters from streams in eastern North America.[336] In this study researchers first examined several different populations of one particular darter species, the tessellated darter (*Etheostoma olmstedi*) (figure 9.6). The rationale was that, if these populations are in the early stages of becoming new species, then we might expect the epigenetic divergence among the populations to be larger than the genetic divergence. This is precisely what they found. As a second step, they then also examined genetic and epigenetic divergence among different species of darters at a broader scale and explored how well genetic versus epigenetic differences can explain the degree of reproductive isolation between these species. Their findings showed that epigenetic distance between species was a significant predictor of reproductive isolation whereas genetic distance was not. Both findings therefore agree with the broad qualitative predictions from theory.

The second study involves five species of the iconic group of birds called Darwin's finches from the Galápagos Islands (figure 9.7).[337] Again, the goal of the study was to explore whether epigenetic or genetic change tends to underlie the differences among the species, and again the findings were broadly in line with theoretical predictions. The five species of finches tended to differ more epigenetically than they did genetically.

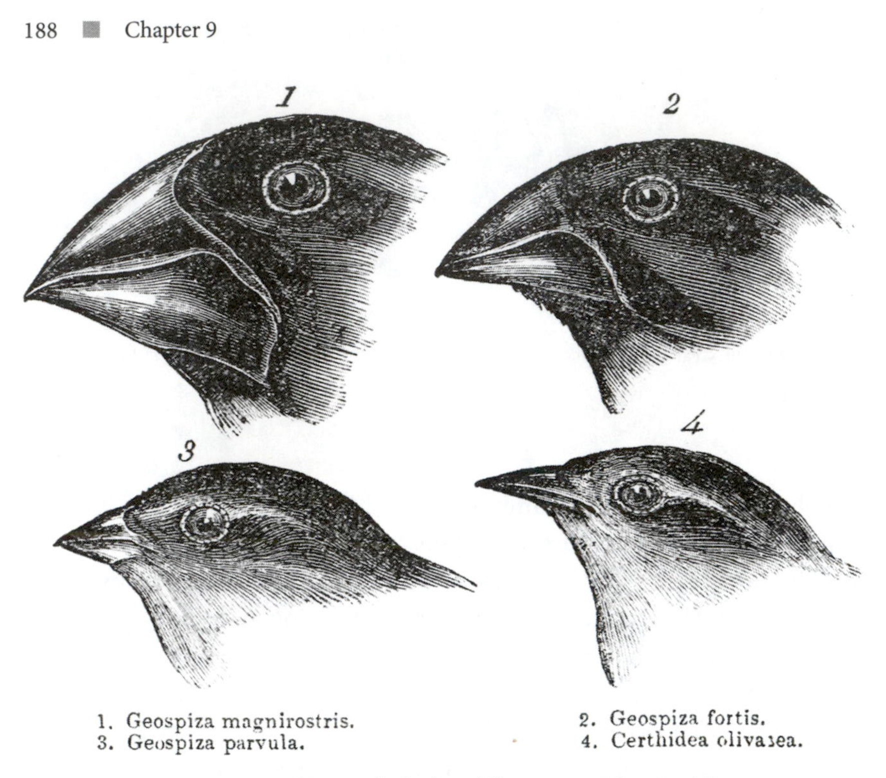

Figure 9.7. Various species of Darwin's finches. (Illustration: John Gould)

Also, the epigenetic differences between species tended to be larger for pairs of species that were more distantly related, whereas this was not consistently the case when considering genetic differences (at least for the type of genetic variation examined in this study—the number of copies of repeated genetic elements present in the genome).

These preliminary studies on the role of epigenetics in the process of speciation are exciting, and they are consistent with the theoretical prediction that epigenetic inheritance allows for peak shifts by initiating evolutionary change, followed later by genetic change. At the same time, however, caution needs to be exercised before reading too much into these results. For example, in neither study is it known if the epigenetic markers studied are actually inherited from one generation to the next. Moreover, the genetic component of the populations analyzed was just one arbitrary part of the broader genome. Consequently, as we noted in previous chapters, it remains possible that the epigenetic differences measured are not heritable but instead are coded for by some other, unmeasured, component of the genome, or that the epigenetic differences are simply plastic responses to the environment. Given these unknowns, a lot of further work must be done on this interesting question.

MAJOR EVOLUTIONARY TRANSITIONS AS EXTENSIONS OF HEREDITY

In their influential book *The Major Transitions in Evolution*, John Maynard Smith and Eörs Szathmáry identified several profound transitions that occurred over the course of evolutionary history. Many of these transitions involved changes in the nature of reproduction, whereby entities that formerly existed as independently reproducing individuals began to exist, experience selection, and reproduce as a unit. This type of transition is seen, for example, in the evolution of the eukaryotic cell through the symbiosis of two very different kinds of prokaryotic organisms (archaea and bacteria), the evolution of multicellular eukaryotes from single-celled eukaryotes, and the evolution of social animals from solitary ancestors. Maynard Smith and Szathmáry also noted that a number of transitions involved the evolution of new forms of heredity, such as the genetic inheritance system in primitive cells, the complex epigenetic system in eukaryotes, and cumulative cultural inheritance in the human lineage.[338]

From our perspective, some interesting developments in the evolution of heredity were also associated with the appearance of the major lineages of multicelled organisms. Some lineages of metazoan animals, such as the Ecdysozoans (arthropods, nematodes, and their relatives) and Chordates (vertebrates and their relatives), evolved a specialized germ line that is sequestered in early development, giving rise to the "Weismann barrier" separating germ line from soma. As we have seen, this barrier is more like a sieve than like a brick wall, but it undoubtedly shields the germ line from some types of environmental influences. (Interestingly, the evolution of eusociality may have added a new layer to this barrier: in eusocial species, the germ line is sequestered inside the body of the reproductive "queen," who is, in turn, sequestered inside the colony.) Other lineages, including animals such as mollusks, annelids, and echinoderms, as well as plants, produce their germ-line tissue at the adult stage.[339] Such organisms therefore lack a true "Weismann barrier," perhaps leaving their germ line exposed to a greater range of environmental influences. However, even in lineages that possess a "Weismann barrier," limitations on nongenetic inheritance imposed by germ-line sequestration have often been compensated by the evolution of new mechanisms of nongenetic inheritance associated with innovations in reproduction. For example, most animals engage in sexual reproduction, which affords opportunities for paternal effects. Likewise, many animals and plants exhibit some form of postfertilization association of parent and offspring, which creates many additional opportunities for maternal (and sometimes paternal) influence on offspring development. The evolution of complex brains and behaviors in some animal lineages also made possible the parent-offspring transmission of behavioral variation. The evolution of complex physiologies, nervous systems, and reproductive modes in multicelled lineages therefore led to a number of important new developments in heredity, such as the appearance of more extensive diversity among lineages in the nature of heredity, and the origin of postfertilization mechanisms of inheritance.

Moreover, as Eva Jablonka and Marion Lamb have emphasized, new mechanisms of inheritance were not only consequences but also causes of evolutionary transitions. The successive elaborations of heredity must have altered the evolutionary potential of the lineages that underwent these transitions. While nongenetic inheritance already occurs in the simplest prokaryotes, the appearance in eukaryotes of new nongenetic mechanisms that follow different rules undoubtedly created new opportunities for nongenetic evolutionary responses as well as more complex interactions between genetic and nongenetic factors.

Such hereditary diversity might have shaped many aspects of macroevolutionary history. For example, Maynard Smith and Szathmáry note that the failure of prokaryotic lineages to evolve multicellularity could relate to differences in cell structure and scope for epigenetic inheritance. Prokaryotes possess a simple epigenetic machinery, and it's possible to imagine how this machinery could support cell-lineage specialization within a multicelled body. However, the far more complex and flexible epigenetic systems that evolved in eukaryotic cells and the potential for virtually unlimited cell-to-cell heredity engendered by these systems represents the most plausible explanation for the evolution of multicellularity in eukaryotes. These epigenetic systems may also have played an essential role in the evolution of sex. We have already noted that variation in epigenetic and cytoplasmic quality may have selected for strategies involving cell fusion, and it's possible that such strategies represented the first steps in the evolution of sexual reproduction. Likewise, as we have seen, nongenetic inheritance could play important roles in the evolution of aging and the evolution of mating systems. And, of course, there can be little doubt that cumulative culture and symbolic communication played a central role in human evolution; the myriad interactions of cultural inheritance with genes may have allowed for the evolution of many unique human traits, including complex theory of mind, and language. Thus, the diversification of nongenetic inheritance mechanisms may have had a profound influence on evolutionary history.

In the next and final chapter, we will turn the spotlight on our own species and explore some of the practical implications of extended heredity.

10

Extended Heredity in Human Life

> Are we no more than passive transmitters of a nature we have received, and which we have no power to modify?
>
> —Francis Galton, "Hereditary Character and Talent," 1865

The article from which this Galton quotation was taken appeared in a high-brow London literary journal in the same year that a monk named Gregor Mendel presented a paper on pea plants to the Brünn Natural History Society in faraway Brno, Habsburg Moravia. No one at the time could have foreseen the roles that Galton's eloquent fulmination on the inheritance of human talent and Mendel's obscure technical report on plant breeding would play in the history of biology, nor their far-reaching consequences for human life. Today, Mendel's contribution is widely known, but the influence of Galton's ideas is underappreciated. Yet, over the decades that followed, while Mendel's work still languished in dusty archives, it was Galton's writings that convinced influential contemporaries that hereditary factors ("nature") were autonomous and independent of the mortal body, while environment and upbringing ("nurture") was less important and could have no effect whatsoever on descendants. The scientific worldview that emerged in the early twentieth century embodied these Galtonian ideas. This view was summed up by the science writer Amram Scheinfeld in his 1939 essay "You and Heredity":

> What this means is that no change that we make in ourselves or that is made in us in our lifetimes, for better or for worse, can be passed on to our children through the process of physical heredity.[340]

Readers who have followed us this far will realize that this view of heredity is contradicted by a great deal of evidence; numerous studies have shown that what happens to us during our lifetimes can influence our descendants. We have devoted most of this book to making the case that such effects should no longer be neglected in evolutionary research, but in this final chapter we will briefly explore the tangible impact of such effects—and the sometimes-tragic consequences of their neglect—for the practical concerns of human beings in the modern world.

HALF BURNT UP AND SHRIVELLED

In 1973, researchers from the University of Washington examined children with peculiar birth defects—characteristic facial and skeletal deformities accompanied by retarded growth, learning disabilities, and impaired motor control—and drew a startling conclusion: the children were victims of heavy alcohol consumption by their mothers during pregnancy. The authors noted that they were unaware of any previous reports of an association between maternal alcoholism and congenital abnormalities in children,[341] and it was only in the 1980s that the US government began to issue warnings to pregnant women about the risks of heavy drinking during pregnancy.[342]

What's most astonishing about the discovery of fetal alcohol spectrum disorders (FASD) in the 1970s is that the link between maternal drinking during pregnancy and developmental abnormalities in children was already known in ancient Greece and Rome and was a prominent social concern in Europe during the eighteenth and nineteenth centuries (figure 10.1).[343] In 1726, the Royal College of Physicians in the United Kingdom cited congenital abnormalities in children in its plea for government controls on the distribution of gin[344] and, a few years later, in a treatise on the evils of alcohol, the English clergyman and social activist Thomas Wilson summarized the expertise of "eminent Physicians in Town" as follows:

> But these Distilled Spiritous Liquors have a more certain ill Effect upon the Children of the Mothers that habituate themselves to the Drinking of them, who come half burnt up and shrivelled into the World . . . If

Figure 10.1. Detail from the etching *Gin Lane* (1751) by the London artist and social critic William Hogarth. The risks of alcohol consumption for pregnant women and their children were already widely known by the early twentieth century. Yet these risks were forgotten for several decades, to be rediscovered only in the 1970s. (© Tate London, 2017)

therefore Child-bearing Women are habituated to strong inflaming Liquors, the little Embryos must and will have a Share.[345]

By the early twentieth century, alcohol's effects on fetal development had been demonstrated in a number of studies on human cohorts. In their review of the historical literature, Rebecca Warner and Henry Rosett concluded that "the physicians and heads of sanitariums who dealt with problem drinkers seem to have had no doubt . . . that parental drinking damaged the offspring."[346] Research on laboratory animals generally pointed to the same conclusion. For example, Charles Stockard and George Papanicolaou[347] of Cornell Medical School studied the effects of alcohol on guinea pigs over several years and, by 1918, concluded that maternal alcohol consumption directly affected developing embryos, while both maternal and paternal alcohol intake appeared to damage the hereditary material ("germ plasm") within eggs and sperm, resulting in a range of developmental abnormalities in offspring and even grand-offspring.[348]

Yet, scientists' views soon underwent a startling transformation—a change so profound that, by the early 1930s, the existence of what we now know as FASD came to be dismissed as a myth and largely forgotten for the next four decades. How could that have happened?

The story of how FASD came to be dismissed by biomedical science is complex: World War I, the enactment of Prohibition laws in the USA, changing social attitudes toward women and drinking, and even the eugenics movement all played a role.[349] Yet, the complete dismissal by scientists on both sides of the Atlantic of a large body of evidence on FASD only makes sense in the context of the prevailing intellectual atmosphere—a scientific revolution-in-progress whose central plank was the belief that Mendelian genes are the sole hereditary factors transmitted from parents to their offspring, and whose defining battle was the purging of all vestiges of "Lamarckism" from biology.

To understand what happened, it's important to realize that, in the early twentieth century, the effects of parental alcohol consumption on children's features were regarded as manifestations of soft inheritance.[350] Some researchers believed that alcohol damaged the "germ plasm" within eggs and sperm and viewed these effects as evidence that the environment can alter hereditary factors in a predictable way, in an apparent violation of the "Weismann barrier." For example, based on their experiments on alcohol's effects in guinea pigs, Stockard and Papanicolau concluded:

> Any strange chemical substance which may find its way into the body fluids will reach the germ cells, [which] may be so modified as to render them incapable of normal development . . . then not only will the generation resulting from the originally modified germ cells be affected, but all future generations arising from this modified germ plasm will likewise be affected.[351]

During Prohibition (1919–33), when alcohol's effects on human health ceased to be a pressing concern for the American public, scientists working on FASD began to place even greater emphasis on the theoretical implications of their studies, and their work came to be seen as a test of the possibility of "Lamarckian" inheritance.[352] For example, in a paper presented in 1923 at the American Philosophical Society's symposium on the inheritance of acquired traits, Frank Hanson of Washington University capped his discussion of alcohol's effects on offspring by concluding that "it appears that after all these years Lamarck's theory is still in the 'pro' and 'con' stage."[353]

But by the 1930s the debate was over, the inheritance of acquired traits having been declared a myth refuted by the new science of Mendelian genetics. As we saw in chapter 3, this consensus was based on shaky evidence and logic, but it was to stand virtually unchallenged for several decades. And, as part of the process of cleansing biology of "Lamarckism," FASD had to be disposed of as well.

Scientists dealt with this problem by reinterpreting the evidence. To reconcile findings from human cohort studies with Mendelian genetics, scientists argued that it was not the effects of parental alcohol consumption per se that resulted in congenital abnormalities in their children; it was simply a case of bad genes causing problems for both generations. In other words, genes present in alcoholic parents as well as their children resulted both in a tendency to alcoholism and in abnormal embryonic development.[354] For example, the editors of the *Journal of the American Medical Association* assured their readers as follows:

> It is an old but unsubstantiated belief that acute alcoholic intoxication of the parents at the time of conception has a detrimental effect on the offspring . . . for instance, that the incidence of mental deficiency is greater among the offspring of drunkards than in the general population. However, it has been also found that drunkards come frequently from families with hereditary moronism, and this heredity accounts for the moronism of the drunkard's offspring rather than germ damage from alcohol.[355]

Or, as Howard Haggard and Elvin Jellinek put it in their 1942 book *Alcohol Explored*, "while alcohol does not make bad stock, many alcoholics come from bad stock."[356] With this ingenious argument, it was no longer necessary to accept that an environmental factor like alcohol could predictably alter the features of offspring, in apparent violation of the "Weismann barrier."

Bad genes could not account for the results of experiments on laboratory animals, such as those by Stockard and Papanicolau. However, given the relatively primitive experimental and analytical techniques available in the early twentieth century (a problem we considered in chapter 3), the experimental evidence was mixed and not straightforward to interpret, so later researchers could simply dismiss it as unreliable.[357] As Amram Scheinfeld put it in 1939, "Certain experiments were reported as proving that drunkenness, and other dangerous habits, could be passed on by heredity. All these 'findings' have since been discredited."[358]

Having thus dealt with the evidence, British and American scientists were for several decades practically unanimous in denying that alcohol could have any effects on development.[359] Haggard and Jellinek summarized the consensus in unambiguous terms:

> The fact is that no acceptable evidence has ever been offered to show that acute alcoholic intoxication has any effect whatsoever on the human germ, or has any influence in altering heredity, or is the cause of any abnormality in the child.[360]

Writing in 1964, just a few years before the rediscovery of FASD, anthropologist Ashley Montagu was even more emphatic:

> It can now be stated categorically, after hundreds of studies covering many years, that no matter how great the amounts of alcohol taken by the mother—or by the father, for that matter—neither the germ cells nor the development of the child will be affected.[361]

But this mid-twentieth-century scientific consensus was horribly wrong. Since the 1970s, the reality of FASD has been demonstrated in hundreds of studies, and maternal alcoholism is recognized today as a leading cause of developmental abnormalities and mental retardation in children.[362] While alcohol's effects may not be transmissible over multiple generations as some early researchers supposed, there is no question of the potential for effects on embryos exposed in the womb. Several studies also support the results of early twentieth-century experiments by providing evidence that paternal drinking can influence offspring via sperm-borne effects,[363] perhaps involving alcohol-induced epigenetic changes.[364] In other words, it turns out that, even if alcohol cannot alter germ-line DNA sequences, it can affect embryonic development in other ways that mid-twentieth-century biologists were simply unaware of. Today, women who are pregnant or trying to conceive are now routinely warned to avoid alcohol (and, someday, similar warnings may be issued to would-be fathers as well).

WITHOUT ADVERSE EFFECT ON MOTHER OR CHILD

No one knows how many preventable cases of FASD occurred as collateral damage from the purge of "Lamarckism" from biology. Indeed, we wonder whether the same mentality might also have contributed to other tragic medical failures, such as the thalidomide debacle. Developed in West Germany in 1954 by the drug company Grünenthal, thalidomide was widely marketed to pregnant women in the late 1950s and early '60s as a sedative and cure for morning sickness. Grünenthal claimed that thalidomide had no side effects whatsoever, while the UK distributor, DCBL, promised that it "can be given with complete safety to pregnant women and nursing mothers, without adverse effect on mother or child." But thalidomide was ultimately shown to have caused severe congenital deformities—missing or shrunken limbs, often accompanied by defects of the nervous system and other parts of the body—in thousands of children in several countries.[365]

Prior to approval for human use, thalidomide's safety was assessed in adult people and rats, but no tests were conducted for risks to developing embryos. The reasons for this oversight are complex. It's clear that companies marketing thalidomide were keen to downplay the potential risks, while governmental authorities that approved thalidomide were guilty of negligence. But the attitudes of doctors and medical researchers must also have played a role. Although a number of studies had already demonstrated the potential for chemicals to cross from mother to fetus via the placenta,[366] it was not widely appreciated in the 1950s that a drug that had little or no visible effect on a woman could nonetheless severely harm a fetus in her womb.[367] The knowledge and beliefs of the doctors who approved and prescribed thalidomide deserve further study, but we suspect that the same Galtonian view of heredity that led to the dismissal of FASD might also have blinded doctors to the potential danger posed by thalidomide. Indeed, one doctor who, at the time, doubted thalidomide's role in the emerging epidemic of congenital abnormalities later acknowledged that his skepticism stemmed from his belief that birth defects must have a genetic basis.[368] The thalidomide debacle led governments to adopt new guidelines for drug testing that required investigation of side effects on developing embryos.[369]

It's reassuring to think that the lessons of the thalidomide tragedy were learned in the 1960s. Yet, much more recent developments cast a disturbing light on thalidomide's legacy. Studies published in the 1990s suggested that the grandchildren of women who took thalidomide might also be disproportionately prone to abnormal development.[370] These reports aroused furious reaction from some doctors and medical scientists, with one pediatrician declaring in the journal *Drug Safety* that "Lamarckism has long since been abandoned by scientists,"[371] and a major London newspaper assuring its readers that "British specialists say it is impossible for a drug to cause a malformation which is then passed on to subsequent generations."[372] These words, written in the late 1990s, leave the reader with the impression that Weismann's tail-clipping experiment was the last word on heredity. To be clear, the problem here is not that the reports of second-generation effects were challenged; the evidence should certainly have been subjected to scrutiny (and, indeed, the jury is still out on thalidomide's second-generation effects). The problem is that such effects were dismissed as *impossible in principle*—that is, precluded by the Galtonian concept of heredity, now reconciled to the possibility of a toxin crossing from mother to embryo, but still adamantly denying the possibility of effects on grandchildren.

NUTRITION IN AND OUT OF THE WOMB

Given the years-long blindness to the striking symptoms of FASD and thalidomide, it need not come as a surprise that more subtle effects took even longer to be recognized. Today, the risks of maternal malnutrition in pregnancy seem self-evident, and few scientists would question the idea that the intrauterine environment can influence fetal growth and development, but as recently as the 1990s, evidence of nutrition-induced maternal effects in humans and other organisms was still a controversial novelty.

Beginning in the 1980s, David Barker and colleagues analyzed data on cohorts of people born in the United Kingdom in the early twentieth century and reported that early-life measures such as birth weight and placenta size predicted health outcomes such as respiratory ailments and heart disease later in life. On the basis of these observations, Barker proposed the "fetal programming hypothesis"—the idea that maternal health and nutrition influence conditions within the womb that, in turn, affect the developing fetus, with potentially life-long consequences for health.[373] Barker's claims initially aroused considerable controversy, with critics (rightly) arguing that the link between maternal nutrition and fetal growth had not been conclusively established, given the powerful confounding variable of familial poverty.[374] However, Barker's key insight has since been confirmed by many studies.[375] Together with Nick Hales, Barker went on to propose the thrifty phenotype hypothesis—the idea that fetal undernutrition results in impaired growth of the β-cells of the islets of Langerhans in the fetal pancreas, leading to impaired capacity to stabilize blood sugar levels throughout life. Hales and Barker reasoned that, if an undernourished fetus is born into a world of superabundant, sugary food, its body will be especially hard-pressed to cope, making undernourishment in the womb followed by overnourishment after birth a recipe for Type II diabetes.[376]

Indeed, overconsumption of calorie-rich food, along with a deficit of physical activity, has become distressingly common in many parts of the world, and Edward Archer at the University of Alabama recently proposed that this combination of diet and lifestyle can give rise to a complex positive feedback loop whereby maternal obesity predisposes children to obesity.[377] Archer argues that maternal obesity leads to oversupply of nutrients to the fetus within the womb, triggering a change in fetal development characterized by excessive proliferation of fat-storing cells (adipocytes) at the expense of other tissues such as skeletal muscle. This results in large, heavy, relatively inactive babies that, because of the overabundance of adipocytes relative to other tissues in their bodies, are physiologically primed to store energy as fat. And, because fat deposition is associated with physiological changes that induce hunger, such children are especially prone to overeating.

But it gets worse. Archer cites studies showing that overweight mothers are especially likely to breastfeed while watching television, perhaps conditioning their infants to associate the flickering images and sounds with food, and many children continue to eat in front of the television as they get older. Of course, poorer families also tend to eat fatty and sugary fast foods and "TV dinners," which are cheap, readily available, and easy to prepare (not to mention scientifically engineered to taste and smell appetizing, and aggressively marketed to children on television). Archer suggests that many families have now reached a tipping point beyond which the cycle of increasing obesity is very difficult to break; even dieting and exercise don't help because a tendency to obesity is built into the body during fetal development, and this combination of physiological and behavioral maternal effects results, in each successive generation, in a greater predisposition to obesity that is even more difficult to counter through lifestyle changes. Some families may also have a genetic predisposition to obesity, or to the maternal effects involved in the obesity-enhancing feedback loop—an interaction of genetic and nongenetic inheritance systems. Archer's hypothesis suggests that nongenetic transmission of a suite of physiological and behavioral traits from mother to daughter is the basis of progressively increasing rates and degrees of obesity, a form of nongenetic evolution manifested in the accelerating obesity epidemic occurring in parts of the world today.

Today, Barker's fetal programming theory, generalized and extended into many new domains, forms the foundation of a burgeoning research program called the developmental origins of health and disease (DOHaD).[378] The basic premise of DOHaD is that many aspects of our physical and mental health as adults are shaped by the environment that we experienced inside the womb or in early childhood; over- or undernutrition, stress, exposure to toxins such as alcohol or nicotine, or mental stimulation, can leave a life-long physiological legacy by altering development. Given that development proceeds through a series of highly stereotyped phases, with specific tissues and organs forming on particular days after conception, and each step providing the essential platform for the next one, DOHaD predicts that the specific consequences of a particular environmental factor will often depend on the timing of fetal exposure (as illustrated by the gestational "timetable" of thalidomide-induced birth defects[379]). Because many aspects of development appear to have evolved to allow the fetus to adjust physiologically to the environment that it can expect to encounter after birth (for example, maternal undernutrition may trigger developmental changes geared to producing a body that can cope with food scarcity), DOHaD also suggests that ill health is especially likely to result when fetal and postnatal environments are mismatched.

DOHaD's predictions can be tested experimentally on laboratory animals such as mice, rats, and rabbits, but, for obvious ethical reasons, studies on human subjects must be correlational, making interpretation far more challenging. For example, as we've already noted, Barker's evidence on fetal programming was originally criticized because low-birth-weight infants also tended to be born into poor families, poor neighborhoods, and disadvantaged ethnic groups. While it's very difficult to fully disentangle such factors in human cohort data, researchers have taken advantage of "natural experiments"—usually involving wartime tragedies—to test DOHaD's predictions.

One of the most famous case studies focuses on the Dutch Hunger Winter, a severe, five-month-long famine that affected the entire population of the western Netherlands near the end of World War II, when Nazi Germany imposed a total embargo on food transport into the region. As a result of the blockade, the energy provided in daily food rations in Amsterdam and surrounding areas dropped from 1,800 to as little as 400 calories per day.[380] By studying the children of women who conceived or were already pregnant during the famine, researchers have been able to learn a great deal about the life-long consequences of maternal nutrition on fetal development. It was found that mothers who starved during the first trimester of pregnancy gave birth to large babies that suffered increased rates of obesity and heart disease as well as impaired cognitive performance later in life, whereas mothers who starved during mid-pregnancy gave birth to children with an increased prevalence of respiratory ailments. By contrast, there was little evidence of ill health in children of mothers who had starved in late pregnancy; such mothers gave birth to small babies that had low rates of obesity throughout life.[381] These findings may seem paradoxical, given that fetal energy demand, and therefore the energy deficit resulting from maternal starvation, increases as pregnancy progresses. However, the striking effects of maternal starvation in the first trimester make sense from a DOHaD perspective; because many aspects of morphology, physiology, and metabolism are "programmed" for life during early stages of fetal development, intrauterine conditions during the first trimester are likely to be of greatest importance, and capacity to compensate for such effects at later stages is limited.

That the poor health of individuals exposed to the famine early in fetal development resulted largely from a mismatch between nutrition levels before and after birth is suggested by a comparison between the Dutch Hunger Winter and another wartime tragedy, the German siege of Leningrad (now Saint Petersburg). After the Dutch famine, energy intake levels returned to normal very quickly, so that children exposed to the famine in the womb were born into a world of plentiful food. In contrast, the Leningrad famine was followed by a long period of restricted caloric intake as the war-ravaged USSR struggled to restore food supplies. Just as the thrifty phenotype hypothesis would predict, the deleterious metabolic effects observed in the children of Dutch Hunger Winter mothers were not observed in the children of Leningrad siege mothers.[382]

While early analyses focused on the children of Dutch Hunger Winter mothers, more recent analyses have also revealed effects on their grandchildren. Researchers found that grandchildren were born shorter and heavier for their size relative to controls.[383] Moreover, these individuals had a higher body mass index as adults (packing an extra 5 kg relative to controls), but, intriguingly, this effect was only observed in individuals whose paternal grandmother had suffered the famine. The hereditary mechanism responsible for these effects is not yet known. Although some analyses have detected epigenetic changes in the children of Dutch Hunger Winter mothers, these children have also been found to be less active and more prone to eat high-fat food than controls, suggesting that behavioral transmission of an unhealthy lifestyle might have contributed to the effect.[384] However, the possibility that effects of maternal malnutrition are at least partly mediated by changes in DNA methylation in embryos is supported by recent studies on a rural population in The Gambia. In this population, food availability and the intake of calories and protein follow a predictable seasonal pattern. Researchers found that children conceived during the rainy season, when food is scarce and farmwork is most intense, differ in methylation patterns at several genes from children conceived during the dry season, when food is abundant, and these DNA methylation patterns persist throughout life.[385] Whatever the mechanism, these findings illustrate the need to take into account the mother's environment and experiences prior to conception and to look beyond a single generation in DOHaD research.

Indeed, many illnesses that run in the family are now recognized as having a nongenetic cause, such as abnormal DNA methylation.[386] For example, disruptions to normal patterns of genomic imprinting (an epigenetic mechanism whereby certain genes are differentially methylated and thus differentially expressed depending on whether they were inherited from the mother or the father) are implicated in Prader-Willi and Angelman syndromes. Other disorders, such as defects of hemoglobin (thalassemias) and certain forms of mental retardation and muscular dystrophy, are also known to involve epigenetic dysregulation. Epimutations transmitted through cell lineages within the body play an important role in some cancers as well. Of course, when transmission occurs from mother to child, it can be challenging to determine whether the altered epigenetic state is transmitted through the germ line, via changes in conditions experienced by the fetus inside the uterus, or even via postnatal factors such as milk or maternal behavior. Although a hereditary illness is just as real regardless of which of these routes of transmission is involved, understanding the mechanism of transmission can inform decisions about prevention and treatment. Such complications are part of the reason why medical researchers are so interested in cases involving transmission from father to child, where at least some channels of nongenetic transmission, like the intrauterine environment, can be ruled out.

SINS OF THE FATHER

DOHaD studies—like Barker's and Archer's hypotheses and analyses of the Dutch Hunger Winter and Leningrad cohort—quite sensibly focus on the role of the mother and the intrauterine environment. But what about fathers? While researchers have now accepted the idea that the mother provides the environment in which her child develops, in most organisms (including humans) the father's contribution to the development of his offspring is typically assumed to consist of a bundle of genes fitted with a tiny outboard motor. But this view is starting to change under the weight of mounting evidence that paternal environment prior to conception can affect children's development and health.[387]

The long-standing neglect of the role of paternal environment stemmed from ignorance of any molecular or physiological mechanisms that could transmit such effects to the fetus, but several candidate mechanisms are now known. A male's environment can modify DNA methylation, chromatin structure, RNA content, or even RNA methylation in the sperm, and the transmission of such epigenetic factors to offspring can modify gene expression during development.[388] In other words, that bundle of paternal genes comes with a set of instructions specifying how those genes will be expressed in development, and those instructions can be damaged or modified by the male's environment, experiences, or senescence. Moreover, sperm make up only a small fraction of an ejaculate, and it looks increasingly as though the seminal fluid itself could play a role in heredity. In mammals, insects, and other internally fertilizing animals, seminal fluid is a veritable chemical soup composed of hundreds of different proteins, RNAs, and other molecules,[389] and it appears that a male's environment can affect the composition of this mixture and thereby influence offspring development.[390] For example, a recent study on the tiny nematode worm *Caenorhabditis elegans* showed that paternal RNA plays a very active role in early embryonic development,[391] and the rich assortment of RNA in human semen has the potential to play a similar role.[392] One potential way that environmental effects on the paternal soma could bypass the "Weismann barrier" and infiltrate the seminal fluid is via the mysterious entities known as extracellular vesicles.[393] As we saw in chapter 4, these tiny, membrane-bound bundles of RNA, protein, and other molecules are released from somatic cells, can be transported via the bodily fluids to the gonads, and could be transported from male to female and egg via the seminal fluid.[394]

In chapters 4 and 5, we outlined experimental evidence on laboratory animals showing that paternal diet, stress, or even sensory experiences can influence the features of offspring. Correlational evidence from human cohorts suggests that similar effects are possible in our own species as well. The most famous example comes from the Överkalix region in northern Sweden. Because of its remote location, the population of Överkalix used to subsist almost completely on locally produced food. Local birth and harvest records allowed researchers to determine how much food was available to people born in Överkalix in the late nineteenth and early twentieth centuries, and to correlate the cycles of dearth and plenty during these people's childhood years with the health of their descendants. This detective work revealed an intriguing pattern: the health and life expectancy of grandchildren was found to depend strongly on the amount of food available to their paternal grandparents as children, and the nature of these effects was sex specific.[395] The researchers found that grandsons' longevity depended on the childhood diet of their paternal grandfather, while granddaughters' longevity depended on the childhood diet of their paternal grandmother. Moreover, these effects were negative: grandparents who lived through periods of food limitation as children had longer-lived grandchildren than grandparents who had abundant food throughout childhood. The fact that these grandparental effects appear to be transmitted through sons but not through daughters undoubtedly holds clues, but the mechanism of transmission remains unknown. An effect of diet on DNA methylation patterns in sperm is one obvious candidate mechanism (consistent with evidence of differences in DNA methylation in sperm of lean and obese men[396]), but changes in sperm-borne RNA or chromatin structure, or even seminal fluid-borne factors, are also plausible candidate mechanisms.

There is also evidence that paternal preconception exposure to particular substances can affect children's physiology and health. In many parts of Asia, a preparation called betel-quid or paan, made from the fruit of the areca palm (*Areca catechu*) wrapped in leaves of the betel vine (*Piper betle*), and often mixed with calcium hydroxide and tobacco, is chewed as a stimulant. A study on the effects of betel-quid chewing in Taiwan showed that men who chew this concoction sire children prone to obesity, high blood pressure, and elevated blood sugar levels, even when the children themselves do not chew betel-quid.[397] Other studies have examined the effects of paternal smoking on children's health, using data from a cohort of children born near Bristol, United Kingdom, in the early 1990s.[398] It was found that fathers who started smoking earlier—and especially those who smoked from childhood—produced sons prone to obesity.

These findings suggest an important paternal dimension to DOHaD that has been largely ignored until very recently.[399] While it is the mother who provides the physical surroundings in which the fetus develops and furnishes the energy and nutrients that enable it to grow, it is becoming clear that fathers can transmit environmental influences to their children alongside their genes.

PSYCHOLOGICAL TRAUMA

In 1939, the Jewish population of Europe totaled nearly ten million people, most of whom lived in the old "pale of settlement" established by Catherine the Great on the western fringes of the Russian Empire.

Over the following six years, as central and eastern Europe fell to Nazi Germany, Hitler's eugenics-inspired genocidal plan was put into effect. By 1945, nearly two-thirds of Europe's Jews had been killed, and many of the survivors carried psychological scars from the horrors they had witnessed.

While the population of Holocaust survivors is rapidly dwindling today, research suggests that the children of survivors carry with them an echo of their parents' psychological trauma. Studies by Rachel Yehuda at the Mount Sinai Hospital in New York City show that Holocaust survivors who suffered from posttraumatic stress disorder (PTSD) have children whose levels of the stress hormone cortisol are low, compared to a control group whose parents did not suffer from PTSD.[400] Since the release of cortisol helps the body and brain to cope with stress, reduced cortisol levels may indicate that these descendants of Holocaust survivors are themselves more vulnerable to trauma. The mechanism whereby parental PTSD influences cortisol levels in children is not yet known. A recent study has identified changes in DNA methylation in both parents and children,[401] but it is not yet clear whether altered epialleles are the mechanism of parent-offspring transmission or merely a downstream consequence of some other factor, such as in utero effects on fetal development or postpartum effects of maternal behavior (indeed, even genetic effects cannot be entirely excluded, since some individuals may have a genetic predisposition to develop PTSD in traumatic circumstances).[402]

The notion that the scars of war and famine disappear completely with the death of the traumatized individuals is therefore contradicted by several lines of evidence: the legacy of the Holocaust may live on in the children of survivors, just as the Dutch famine continues to afflict the descendants of Hunger Winter mothers. But, of course, each and every human being has ancestors who experienced the physical and psychological stresses of violence, famine, epidemics, and other traumatic events, and millions of people are living through such trials today. We still know very little about how these experiences might affect their descendants.[403]

The studies outlined above suggest that some effects of physical and psychological trauma might be transmitted to children and grandchildren, but much remains unknown. How many generations must pass before the effects of ancestral trauma are extinguished completely? Can trauma that recurs over multiple generations have cumulative effects? Or, conversely, do repeated traumatic episodes over several generations induce physical or psychological defenses that offer some protection against lasting effects of trauma? Can multiple forms of stress, such as psychological trauma and malnutrition, interact in ways that are worse than the sum of their parts?

The case of the Leningrad siege also suggests that the severity of such effects can be strongly context dependent. Factors such as food abundance, socioeconomic status, and social integration might either dampen or reinforce the effects of ancestral trauma. Indeed, the potential for ancestors' stressful experiences to exacerbate effects of stresses experienced by their descendants has been demonstrated experimentally in rats. Males whose grandparents had been exposed to the estrogen-mimicking chemical vinclozolin in the womb were more strongly affected by the stress of being restrained, showing greater anxiety when placed into an open, novel environment. The same study also found that descendants of vinclozolin-exposed animals were more prone to weight gain.[404] The potential for such interactions between ancestral trauma and the environment of descendants remains poorly understood in humans.

Perhaps most challenging of all is to understand the consequences of ancestral trauma at the population level. If enough individuals in a population are affected, horizontal influences might result in emergent effects that reinforce and accentuate the legacy of ancestral trauma. Populations with a history of violence, famine, or persecution might therefore suffer not only from the individual-level effects of ancestral trauma but also from aggregate effects at the population level that act to reinforce these effects. A population in which many individuals were exposed to famine in utero might develop social norms or practices (like reduced levels of physical activity, or increased consumption of high-fat foods) that reinforce and perpetuate the detrimental effects of maternal starvation on health. Similarly, in a population that includes many individuals descended from psychologically traumatized ancestors, increased individual susceptibility to stress might be reinforced by interactions among individuals. In this way, the effects of ancestral trauma might continue to ripple through a population for generations.

CHASING THE FITNESS MOUNTAIN

Our world has been transformed beyond recognition over the past couple of centuries, and the pace of change is increasing at a breathtaking rate. Based on biological, atmospheric, and geochemical evidence, some scientists have formally proposed a new geological epoch—the Anthropocene—with a boundary in the mid-twentieth century.[405] This new epoch is characterized by a radically altered biosphere, dominated by one species of cantankerous ape whose prodigious population subsists on vast monocultures of plant and animal symbionts. Its activities are bringing about rapidly rising atmospheric and ocean temperatures driven by the release of greenhouse gasses, and generating megatons of agricultural and industrial waste products that accumulate in the oceans and in the soil, forming a new geological layer as distinctive as the layer of deadly dust from the asteroid that ended the age of dinosaurs. Anthropogenic climate change also appears to be driving increasingly rapid and unpredictable oscillations in temperature and precipitation, exemplified by the El Niño climate oscillation in the southern Pacific Ocean.[406] These changes are leading to rapid degradation and loss of natural habitats and pushing numerous species to the brink of extinction.[407]

Understanding how species adapt to rapidly changing and increasingly unpredictable environments is therefore no longer just an esoteric goal of basic science; such an understanding is needed to predict the fate of our fellow creatures, and perhaps ourselves as well. This is inherently a question about rapid evolution, and therefore precisely the kind of situation in which extended heredity could make a big difference.

Researchers began to ponder the role of nongenetic inheritance in population survival and adaptation in the 1990s. These early efforts, spearheaded by Eva Jablonka in Israel, Csaba Pál in Hungary, and others, showed that nongenetic inheritance could act as an evolutionary rapid-response mechanism as well as a kind of medium-term phenotypic memory, helping populations persist in rapidly changing environments whose demands they could not keep up with by the slow process of genetic change.[408] For example, parts of Australia are subject to cycles of drought lasting several years. The oscillations between wet and dry conditions occur too rapidly to allow for substantive genetic adaptation, yet slowly enough that the conditions experienced by parents generally predict the conditions that their offspring will experience. Such conditions may select for "anticipatory" parental effects, a form of transgenerational plasticity whereby parents adjust the phenotype of their offspring for the anticipated conditions—a prediction recently confirmed experimentally.[409] But even if environmental fluctuations are too unpredictable or too complex and multidimensional to allow for the evolution of anticipatory parental effects, selection on random nongenetic variation (such as heritable epigenetic marks) could allow populations to track changing fitness optima more rapidly than they could through genetic change—a possibility that we illustrated in chapter 7 using the analogy of the fleet-footed nongenetic mountaineer.

Nongenetic inheritance, in its many forms, can therefore play a key role in allowing populations to adapt to rapidly changing environments, and this means that understanding the potential for threatened populations to undergo rapid adaptation requires not only a knowledge of their genetic variability and capacity for direct developmental response to environment (plasticity) but also an in-depth understanding of the role of nongenetic inheritance in their heredity. This will involve asking a new kind of question about heredity—a question about taxonomic variation in inheritance mechanisms. For example, given their very different biology, we may expect nongenetic inheritance to contribute very differently to adaptation in corals, Galápagos tortoises, and bonobos. Coral polyps are colonial, externally fertilizing organisms that lack a brain but have a potentially extensive capacity for transmission of epigenetic, cytoplasmic, and symbiotic variation. Tortoises might possess a capacity for nongenetic transmission of diet choice, and perhaps a potential for transmission of nest-site choice as well as the syndrome of traits influenced by nest temperature. Bonobos are highly social creatures with a capacity for behavioral innovation and cultural transmission. Efforts to save each of these organisms from extinction could therefore be furthered by an in-depth understanding of the types of nongenetic inheritance mechanisms that operate in each species and the potential for these nongenetic factors to respond to natural selection and to interact with genes.

Indeed, similar considerations apply to our own species and our capacity to adapt to the challenges of modern life. For people in troubled parts of the world, modern life often means overcrowding, poor sanitation, infectious diseases, and violence. For those privileged to live in peaceful, prosperous societies, the greatest challenges may be posed by such "first world" problems as an overabundance of calorie-rich foods, a sedentary lifestyle, the psychological stresses of big-city life, and the diseases of aging.[410] Some of these challenges appear to result from a mismatch between people's inherited traits (such as skeletal structure, digestive systems, metabolism, and psychological coping mechanisms) and the environments that they have constructed for themselves. This mismatch is typically construed as a problem of genetic adaptation.[411] Yet, despite the undeniable role of genes, people also carry a nongenetic legacy in their bodies, and part of the mismatch may reflect this legacy. Analyses of human cohorts, such as those from Överkalix, the Dutch Hunger Winter, and the Leningrad siege, as well as experimental studies on rodents, suggest that some of the health problems experienced by people today may result from the lingering nongenetic consequences of a rapid transition in diet, lifestyle, and exposure to hormone-mimicking chemicals that their parents or grandparents underwent. Some of these nongenetic changes may be self-perpetuating. For example, as we already noted, the obesity epidemic currently afflicting parts of the world may be an example of nongenetic evolution driven by a combination of epigenetic, somatic, and behavioral factors. Such nongenetic changes are also likely to interact with genetic factors and drive genetic change, as they probably have throughout human evolution.[412] A better understanding of extended heredity is needed to recognize these phenomena and mitigate their effects.

MICROBIAL WARFARE AND RESISTANCE

Although we often think of rapid evolutionary change as being desirable in a world threatened by climate change, it can also be the source of untold human misery. Perhaps the clearest example involves bacterial infections. Prior to the large-scale production and use of antibiotics in the middle of the last century, even minor scrapes and abrasions had the potential to become life-threatening illnesses as a result of bacterial infections.[413] Likewise, the widespread use of countless surgical procedures wouldn't be possible today if it weren't for the ready availability of antibiotics as a means of preventing and curing infections. But the effectiveness of these so-called miracle drugs is currently being undermined by rapid evolution, as their very use drives the emergence of drug resistance in numerous bacterial species.

A glance at the headlines in recent years would lead one to believe that the evolution of drug resistance is a phenomenon that caught nearly everyone off guard, but closer examination shows that clear signs of the impending problem were evident early on. Indeed, in his 1945 Noble Prize lecture about the discovery of penicillin, Sir Alexander Fleming cautioned that when using penicillin, "the ignorant man may easily underdose himself and by exposing his microbes to non-lethal quantities of the drug make them resistant."[414] Perhaps even more intriguingly, the early warning signs of drug resistance also clearly showed that nongenetic forms of variation are likely to blame.

Many people are well acquainted with the story of Fleming's discovery of penicillin when, in 1928, he returned from vacation to discover a mold growing on his petri dishes and recognized that this mold had the ability to kill the surrounding bacteria. This killing ability was due to a substance secreted by the mold, which he named penicillin. In fact, many microorganisms have the ability to produce such toxins as a means of eliminating their competitors. The success of much of modern medicine now depends on our ability to harness this intricate form of microbial warfare.

However, in 1944, one year prior to Fleming's Nobel Prize, a short paper raising concerns about penicillin resistance was published in the medical journal *The Lancet* by an Irish contemporary named Joseph Bigger. In this paper Bigger showed that when bacterial cultures are exposed to penicillin, a small fraction of bacteria often survive. Today we might hypothesize that these survivors are genetically resistant to the drug, and Bigger reasoned similarly, arguing that if the ability to withstand penicillin was a fixed property of the bacteria, then "it is probable that their descendants would also possess . . . resistance."[415] To test this idea Bigger therefore cultured the survivors and then exposed their descendants to penicillin. To his surprise, rather than withstanding the drug, the descendants of the original survivors were just as susceptible to penicillin as was the original population. But, just like the original population, a small fraction of these bacteria again survived the drug. In today's language Bigger therefore concluded that the ability to withstand penicillin must be a highly mutable, phenotypic property of the bacteria rather than a genetically encoded trait. Otherwise we would not expect to see such rapid reversion to the drug susceptible state from the original surviving population. He therefore coined the term "persisters" in reference to these surviving cells in order to differentiate them from bacteria that are (genetically) drug resistant.

Bigger's research did not attract much attention at the time, and he went on to pursue other interests, including being elected to the Irish senate in 1947. In recent years, though, there has been an enormous resurgence of interest in his work. For example, we now know that persister cells are able to withstand drug treatment by going dormant instead of reproducing.[416] Furthermore, a cell's propensity to become dormant appears to be affected by the amount of a certain protein molecule in the cell—once the level exceeds a threshold, dormancy occurs.[417] Now just as with the Venter cell from chapter 7, nongenetic factors like dormancy-causing proteins will likely be transmitted with the cytoplasm during cell division, and these dormancy proteins will then be diluted each generation as well as supplemented by newly synthesized, genetically encoded protein. Thus, the propensity of a cell to become dormant and so withstand drug treatment might be predicted to evolve over time as a result of an interaction between genetic and nongenetic inheritance. If this speculation is correct, then a complete understanding of the evolutionary consequences of antibiotics will only be possible through the use of a framework involving extended heredity.

The resurgence of interest in persister cells today stems, in large part, from the growing recognition that many instances of medical-treatment failure are likely due to persister cells. But persister cells are not the only mechanism through which nongenetic inheritance can affect the evolution of drug resistance. Recent research has also shown that asymmetric partitioning of certain membrane complexes during bacterial cell division can also create phenotypic heterogeneity among descendant bacteria, resulting in some being drug resistant while others are drug sensitive, despite being genetically identical.[418] There is also a growing realization that much of the drug resistance seen in the chemotherapeutic treatment of cancers might also be due to nongenetic inheritance.[419]

All of these results again illustrate that the potential importance of extended heredity reaches well beyond academic pursuits and might have serious implications for how we treat a variety of human diseases.

FROM MICROPIGS TO WOODY GUTHRIE

Nongenetic inheritance also has profound implications for the controversial and hotly debated topic of genetic engineering. The genetic manipulation of organisms, be it for the development of enhanced or sustainable food sources, or for the treatment of disease, has always aroused strong sentiments both in favor of and opposed to the technology. A recently developed biomolecular tool called CRISPR-Cas now allows extremely precise genetic manipulation of many different organisms (Box 10.1), and so has thrust these issues back into the spotlight.

Box 10.1. CRISPR-Cas

In 1987 Japanese researcher Atsuo Nakata discovered a peculiar pattern of short repeated DNA sequences in bacteria.[420] These repeated DNA sequences eventually came to be called (C)lustered (R)egularly (I)nterspersed (S)hort (P)alindromic (R)epeats, or CRISPR. Proteins coded for by genes associated with these repeats came to be called "Cas" proteins (and genes), standing for (C)RISPR (as)sociated proteins. Further research over the next two decades showed that CRISPR-Cas occurs in a variety of bacteria and began to reveal the function of these repetitive sequences.[421] It was eventually established experimentally that CRISPR-Cas is a potent bacterial immune system that maintains a memory of previous infections by viruses.[422] When a virus infects a bacterium, it releases its genetic material into the cell and uses the cell's machinery to replicate. During this process, some Cas proteins take pieces of the viral genome and insert them between repeated DNA segments of the bacterial genome. These spacers contain a genetic fingerprint of the previous infections sustained by that line of bacteria.[423] If the bacteria are later infected with a virus whose genome matches one of the CRISPR spacers, then other Cas proteins cut the foreign genetic material precisely at the sequence location specified by the spacer, thereby preventing it from replicating.

Scientists quickly realized that this evolutionary marvel of an immune system could be leveraged for use in genetic engineering.[424] For example, if one wanted to make a cut at a specific location of a DNA sequence, all one had to do was give the CRISPR-Cas system a spacer with the target sequence and insert this into the cell. CRISPR-Cas would then cut the DNA at the desired spot. After such a cut, the cell typically tries to repair the DNA, but errors usually occur, and so the gene at that location is inactivated. Further, if two cuts are made, then a piece of DNA can be removed. If a new segment of DNA with some desirable properties is also introduced at the same time, then as the cell tries to repair the DNA it is tricked into integrating this new segment of DNA instead. The end result is a cell that has a particular allele removed and replaced with an alternative.[425]

The sheer power of the CRISPR-Cas technology is best illustrated by considering the breadth of applications that are already developed or that are being planned. A recent article[426] in the journal *Nature*, "The CRISPR Zoo," discussed genetically engineering honeybees to be obsessive cleaners in order to prevent disease in honeybee colonies, engineering egg proteins to reduce the incidence of allergic reactions, and enhancing the genetic makeup of endangered Indian elephants with cold-tolerance genes taken from the now extinct woolly mammoth. There is even a genomics company using CRISPR to create 15 kg "micropigs" as pets, complete with plans to allow the choice of different, customizable, coat patterns. But perhaps one of the most potentially alarming applications to date is the creation of viruses containing the CRISPR-Cas technology that can cause respiratory infections in mice, and that genetically modify mouse lung cells to develop cancer.[427] The purpose of this work was to provide an effective means of creating mice suitable for studying lung cancer, but it is not hard to imagine how similar approaches might be used for more nefarious purposes. And the breakneck pace at which this field is developing means that this list will almost certainly be out of date by the time you are reading these words.

Another significant potential use of the CRISPR-Cas technology is for developing and enhancing sustainable food sources. This too has generated controversy, in part because we don't fully understand the consequences of altering our food sources in this way. One side of the argument holds that genetic engineering is really no different from selective breeding, a practice that we have been using to enhance our food ever since the beginnings of animal and plant domestication. The other side of the argument holds that we do not yet know whether direct genetic modification of an organism is effectively equivalent to selective breeding. For example, if we could directly modify a tomato gene to enhance flavor, would the resulting strain of tomato be equivalent to one that would be obtained if we instead painstakingly breed only those tomatoes with the desired flavor over multiple generations? If extended heredity really does play a significant role in evolution, then the direct genetic modification of an organism would *not* be equivalent to selective breeding. Selectively breeding tomatoes for a desired flavor could entail both nongenetic and genetic alterations whereas only the latter would be incorporated with direct genetic modification. This difference could be consequential for food safety. More generally, the reality of extended heredity means that the end product of evolution by natural or artificial selection might well be different from the end product of direct genetic modification.

Perhaps the most significant potential use of the CRISPR-Cas technology, however, is for the treatment of genetic disease in humans. For example, if we understood the genetic basis of a disease then we might use the technology to alter the genes in question so that they no longer cause disease. Not surprisingly, these ideas have also engendered debate, but most of this debate centers on the question of whether we should allow the genetic modification of the human germ line or the modification of human embryos very early in development. Although such modifications could alter the health of the individual, they will also be passed on to future generations and we currently have little understanding of what ramifications this might have. On top of that, although fixing an individual's faulty genetic makeup might be desirable for that individual, do we really have the right to alter the genome of future people who have yet to be born without their consent? And where do we draw the line between what is considered a genetic fault versus acceptable genetic variation? There is a sense in which the human germ line is part of our collective commons. Might we have to safeguard this gene pool as a means of preventing a genetic tragedy of the commons?[428]

This might all sound like science fiction, but CRISPR-Cas really is bringing such fiction to reality. The first attempts at curing human genetic disease will likely involve diseases having a relatively simple genetic cause. One such possibility is Huntington's disease, a genetically inherited disorder that results in neurodegeneration and eventual death. Most cases are caused by a dominant allele at a single locus, meaning that children of an afflicted parent have a 50 percent chance of themselves acquiring the disease. Woody Guthrie,[429] an American folk musician from the Great Depression era, is probably the best-known person to fall victim to the disease. Guthrie had a prolific and very influential musical career despite dying at the relatively young age of fifty-five. Before his death he fathered several children, two of whom also died from the disease. This is perhaps the closest we are likely to get to a clear case for using CRISPR-Cas to modify germ-line cells, but even here there is certainly not a consensus of opinion. Nevertheless, research in this area is already progressing, and government policy makers are struggling to keep up with the breathtaking pace of biotechnological change.

In 2015, the National Institutes of Health (USA) declared that "NIH will not fund any use of gene-editing technologies in human embryos,"[430] and, later that year, a joint meeting of the National Academy of Sciences of the United States, the Chinese Academy of Sciences, and the Royal Society of London recommended a moratorium on deliberate changes in human germ-line DNA sequences because such changes could be passed on to descendants.[431] The United Kingdom's Human Fertilisation and Embryology Authority has since allowed DNA editing in human embryos for research purposes only, expressly forbidding the implantation of such genetically modified embryos into a woman's uterus.[432] Deliberate changes to human germ-line DNA are understandably unpalatable to many people because such interventions are seen as a modern-day form of eugenics.[433] But while scientists and policy makers recoil from this Galtonian idea, Galton's other legacy manifests itself in the scant attention paid to the possibility of heritable nongenetic changes. For example, even though it's now clear that various hormone-mimicking substances and other chemicals can affect embryonic development, there has been little effort to regulate the use of such substances in household products, except in cases where the effects are obvious and immediate. This contrast in attitudes to interventions in genetically and nongenetically transmitted human traits seems difficult to justify on rational grounds. For those concerned about the health of their children and grandchildren, rather than merely about the abstract notion of the "future of the human gene pool," the potential for environmental induction of heritable disease should be a prime concern.

TOWARD A POST-GALTONIAN BIOLOGY

Today, Galton's concept of heredity, embodied in his imagined chain of embryos whose features reflect only their unalterable, inherited "nature," has been undermined by a great deal of research, and its spell is starting to weaken. But his ideas still hold sway in the popular imagination, and their echo continues to reverberate in the halls of science and medicine. Just replace "nature" with "genes" and it's easy to recognize Galton's legacy in modern newspaper articles and popular books; in modern language, this is the implicit idea that genes more or less fully determine our features and the features of our descendants. This notion instills a sense of powerlessness and, at the same time, promotes belief in a kind of genetic redemption. After all, if genes are destiny, then there's no point in fighting their effects. And if the consequences of bad life choices like smoking, junk food, and physical inactivity are limited to our own bodies, then we need not feel guilty for allowing our sins to be visited upon our innocent children because no matter what we do, our children's features will be determined by a random genetic lottery. By now, we hope to have convinced readers that such genetic fatalism is at odds with the evidence. Genes are very important, but it's clear that a variety of environmental factors and experiences—including those that depend on our active choices—can have consequences not only for ourselves but also for our descendants, and such nongenetic effects are especially important on the scale of one or two generations.

Aspects of the modern environment and lifestyle could have dire consequences for future generations. As we saw in chapter 5, some of the artificial chemicals that industry generates and releases into the environment—particularly the myriad hormone-mimicking substances found in pesticides and plastics, even trace quantities of which can disrupt the fine-tuning of the body's biochemical signals—can have disastrous effects on developing embryos and engender transgenerational effects that persist for multiple generations. The list of chemical culprits is long and growing.[434] Some of these chemicals are easily avoidable, at least in the most harmful contexts. Baby bottles made of polycarbonate plastic (which releases bisphenol A, an estrogen-mimicking chemical that disrupts development) were banned a few years ago in some countries after research linked bisphenol A to developmental abnormalities in monkeys and rodents. Although polycarbonate baby bottles continue to be sold in many parts of the world, well-informed parents can simply avoid such bottles. Yet, given the vast variety of different plastics and other chemicals incorporated into everyday products—toothbrushes, kitchen utensils, cleaners, furniture, toys, pens, food packaging—it's virtually impossible for even well-informed individual consumers to avoid the full range of potentially harmful substances. Many industrial chemicals are now present at detectable levels in water, soil, and agricultural products.[435] Even ocean sediments are permeated by tiny particles of plastic that end up in the stomachs of fish, crustaceans, and mollusks, and ultimately find their way to your dinner table.[436] Literally nothing is known about the potential for most of these chemicals to influence germ cells, epigenomes, or the intrauterine environment. Unfortunately, unless their effects are immediate, obvious, and severe, such effects are extremely difficult to detect through casual observation or even detailed data on large cohorts. Experiments on laboratory animals are the only reliable means to test for such effects. Governments must fund this research and enact legislation to protect the public from substances that could harm future generations.

Given the vast number of chemicals and blends of chemicals used today, such testing will be difficult and expensive. However, as knowledge of the biochemistry of developmental processes and signaling systems improves, it may ultimately be possible to better predict effects and prioritize research on those substances that are most likely to cause harm. In the past, such effects were typically discovered only after many cases of severe deformity came to light, because scientists and policy makers simply failed to recognize the need to test for effects on descendants. Today, given the tragic history of FASD and thalidomide, and our much-improved understanding of the potential for hereditary effects of environment, there is no longer any excuse for such myopia. Similarly, there's a need for greater public awareness of the potential for some plastics and other toxins, as well as smoking, diet, and other lifestyle choices, to affect future generations.

Stepping back to survey the current state of knowledge in the study of extended heredity, we are reminded of genetics in the 1920s or molecular biology in the 1950s. We know just enough to fathom the depths of our ignorance and to recognize the challenges that lie ahead. But one conclusion that is already beyond reasonable doubt is that the Galtonian assumptions that have shaped both empirical and theoretical research for nearly a century are violated in many contexts, and this means that biology has exciting times before it. Empirical researchers will be busy for many years exploring the mechanisms of nongenetic inheritance, observing its ecological effects, and establishing its evolutionary consequences. This work will require developing new tools and devising ingenious experiments. Theoreticians have the equally important task of clarifying ideas and generating predictions. And on a practical level, in medicine and public health, it is now equally clear that we need not be "passive transmitters of a nature we have received," because our life experiences play a nontrivial role in shaping the hereditary "nature" that we transmit to our children.

ACKNOWLEDGMENTS

This book could not have been written without the advice, support, and encouragement of many people. Our ideas were shaped by many hours of discussions and correspondence with colleagues, especially Liran Carmel, Anne Charmantier, Stephen Chenoweth, Vincent Colot, Angela Crean, Jennifer Cropley, Étienne Danchin, Damian Dowling, Thomas Flatt, Douglas Futuyma, Lilach Hadany, Lára Hallsson, Edith Heard, Eva Jablonka, Frank Johannes, Dustin Marshall, Shinichi Nakagawa, Daniel Noble, Stewart Plaistow, William Sherwin, Lisa Schwanz, Hamish Spencer, Catherine Suter, Tobias Uller, Jason Wolf, and Carl Zimmer. Liran Carmel kindly shared his unpublished research and provided comments on parts of the manuscript. Angela Crean read multiple drafts and made many helpful suggestions. Shinichi Nakagawa read and discussed each chapter with his students and identified confusing passages. We have learned just as much from former and current students, especially Margo Adler (who read the entire manuscript and gave encouraging feedback), Nathan Burke, Elizabeth Cassidy, Amy Hooper, Erin Macartney, Aidan Runagall-McNaull, Foteini Spagopoulou, and Zachariah Wylde. Three anonymous reviewers made many insightful and constructive suggestions. Their efforts greatly improved this book, but we take full responsibility for any remaining errors, omissions, and bad writing.

Alison Kalett, executive editor for Biology and Neuroscience at Princeton University Press, provided encouragement and helpful advice over several years of planning and writing. Editorial assistant Lauren

Bucca, production editor Jill Harris, and other technical staff at PUP helped with the many procedural steps involved in getting the book into production, and illustrations manager Dimitri Karetnikov helped to redesign figures. Dawn Hall expertly copyedited the manuscript, ruthlessly expunging awkward wording, stray hyphens, and Canadian spelling. Virginia Ling compiled the index.

We thank Ian Tattersall, Kate West (Head of Library) and Sarah Wilmot (Outreach Curator and Science historian) of the John Innes Centre at Norwich Research Park, Jennifer Cropley, Christian Laforsch and Andy Gardner for providing images and/or granting us permission to reproduce them in the book.

Last but not least, we are grateful to our families for their forbearance and support.

NOTES

1 Our position is also distinct from the "extended evolutionary synthesis" (EES), which represents a broader challenge to established evolutionary ideas, including the role of natural selection in adaptive evolution. While the EES encompasses nongenetic forms of inheritance, our concept of extended heredity diverges in important ways from the views espoused by some proponents of the EES. We will outline some of the differences in perspective in chapter 8.

2 Several books have examined the implications of cultural evolution, including R. Boyd and P. J. Richerson, *Culture and the Evolutionary Process*; L. L. Cavalli-Sforza and M. W. Feldman, *Cultural Transmission and Evolution: A Quantitative Approach*; A. Mesoudi, *Cultural Evolution: How Darwinian Theory Can Explain Human Culture and Synthesize the Social Sciences*; P. J. Richerson and R. Boyd, *Not by Genes Alone: How Culture Transforms Human Evolution*. Recent discoveries about the molecular mechanisms of gene regulation are chronicled in two recent books on epigenetics: N. Carey, *The Epigenetics Revolution: How Modern Biology Is Rewriting Our Understanding of Genetics, Disease, and Inheritance*; R. C. Francis, *Epigenetics: How Environment Shapes Our Genes*. The potential evolutionary implications of the complex interactions between organisms and their environment are explored in M. J. West-Eberhard, *Developmental Plasticity and Evolution*; S. E. Sultan, *Organism and Environment*; F. J. Odling-Smee, *Niche Construction*; A. P. Hendry, *Eco-Evolutionary Dynamics*.

3 This quote was found on the office blackboard of the noted physicist and iconoclast Richard Feynman after his death in 1988. James Gleick's book (J. Gleick, *Genius: The Life and Science of Richard Feynman*) provides an authoritative and extremely engaging account of the life and times of Feynman.

4 Venter's group actually carried out two different experiments. First, they extracted the genome from bacterial species *A* and inserted it into bacterial species *B* (C. Lartigue et al., "Genome Transplantation in Bacteria: Changing One Species to Another"). Later, they synthesized the genome of bacterial species *A* and inserted it into bacterial species *B* (D. G. Gibson et al., "Creation of a Bacterial Cell Controlled by a Chemically Synthesized Genome"). It remains unclear whether the chimeric bacteria ever attempted to email Craig Venter.

5 Evolution occurs when natural or artificial selection acts on heritable variation. A variant (e.g., long legs) that is favored by selection (in other words, consistently confers greater-than-average reproductive success) will be represented in a larger proportion of individuals in the following generation. If long-legged parents beget long-legged offspring, and if long-legged parents produce more offspring than the average for the population, then mean leg length will increase from one generation to the next.

6 For a discussion of the enormous challenges involved in the creation of a fully artificial cell, see A. B. Chetverin, "Can a Cell Be Assembled from Its Constituents?"

7 In Venter's experiment, it's thought to have taken dozens of generations of cell division for the artificial genome to convert the features of the host cell lineage into those of the genome-donor species, and such conversion was verified for only a few of the cell's features in a tiny fraction of the chimeric cells. We wonder whether the conversion was total. Some features of the cytoplasm, cell membrane, and epigenome are self-regenerating. Moreover, even if the cell lineage created in Venter's experiment did not retain any features of the cytoplasm-donor bacteria, this could be a highly unusual outcome. The experiment involved billions of cells, only a few of which apparently came to exclusively resemble the genome-donor type in their gene expression profile and proteome. It is possible that the self-regenerating features of the cytoplasm were eliminated only in these extremely rare cases. We will consider Venter's experiment in greater depth in chapter 7.

8 For a very clear exposition of this controversy, see J. Sapp, "Cytoplasmic Heretics."

9 F.V.R. Rozzi and J. M. Bermudez de Castro, "Surprisingly Rapid Growth in Neanderthals"; T. M. Smith et al., "Dental Evidence for Ontogenetic Differences between Modern Humans and Neanderthals."

10 P. Villa and W. Roebroeks, "Neandertal Demise: An Archaeological Analysis of the Modern Human Superiority Complex."

11 D. Gokhman et al., "Reconstructing the DNA Methylation Maps of the Neandertal and the Denisovan."

12 See P. Dominguez-Salas et al., "Maternal Nutrition at Conception Modulates DNA Methylation of Human Metastable Epialleles"; R. A. Waterland et al., "Season of Conception in Rural Gambia Affects DNA Methylation at Putative Human Metastable Epialleles"; D. Gokhman, A. Malul, and L. Carmel, "Inferring Past Environments from Ancient Epigenomes."

13 The "missing heritability" problem reflects the fact that genome-wide association studies have thus far failed to identify genes whose combined effects can account for the observed heritability of many traits, ranging from diseases that "run in the family" to strongly heritable traits such as human height (T. A. Manolio et al., "Finding the Missing Heritability of Complex Diseases"). In other words, although relatives tend to exhibit similar phenotypes for these traits, relatives with similar phenotypes tend to have few genetic alleles in common, and it is therefore not clear what the trait's genetic basis might be. Missing heritability could result from complex interactions between genes (epistasis), since such interactions are difficult to take into account in genome-wide association studies. Missing heritability could also arise if some of the heritable variation for the trait is nongenetic, especially if the nongenetic variation is induced by the environment (R. E. Furrow, F. B. Christiansen, and M. W. Feldman, "Environment-Sensitive Epigenetics and the Heritability of Complex Diseases").

14 We revisit this famous analogy and examine its logic in chapter 3.

15 Biological lineages and populations usually consist of similar but nonidentical beings. Such variation defines what Peter Godfrey-Smith has called the "Darwinian population"—a collection of beings that can transmit their individual variation to their descendants (P. Godfrey-Smith, *Darwinian Populations and Natural Selection*). Darwin was the first (along with A. R. Wallace) to recognize that this kind of population can respond to natural selection and undergo evolutionary change. By contrast, some inanimate objects, like crystals, appear to reproduce themselves but do not constitute a Darwinian population because they lack significant variation and heredity.

16 The fact that all cells come from preexisting cells was originally noted in the middle of the nineteenth century by Robert Remak, although it is often associated with the work of his more famous contemporary Rudolf Virchow.

17 See D. L. Nanney, "Cortical Patterns in Cellular Morphogenesis."

18 The resurgence of interest in the role of nongenetic factors in heredity and evolution was spearheaded by Eva Jablonka and Marion Lamb's 1995 book *Epigenetic Inheritance and Evolution*.

19 See, for example, R. Bonduriansky and T. Day, "Nongenetic Inheritance and Its Evolutionary Implications"; E. Danchin et al., "Beyond DNA: Integrating Inclusive Inheritance into an Extended Theory of Evolution"; K. N. Laland et al., "The Extended Evolutionary Synthesis: Its Structure, Assumptions, and Predictions."

20 For example, individuals who carry mutations for the growth hormone receptor, such as "Laron dwarfs," exhibit greatly reduced ability to secrete insulin-like growth factor 1 (see R. K. Junnila et al., "The GH/IGF-1 Axis in Ageing and Longevity"; Z. Laron, "Laron Syndrome [Primary Growth Hormone Resistance or Insensitivity]: The Personal Experience 1958–2003").

21 Ancient DNA can be sequenced to determine the type or organism from which it came and compare its genes to those of extant species. The distribution of ancient DNA fragments in the environment can even provide information about the ecology of ancient organisms.

22 For example, a parent with an eye color gene variant ("allele") whose expression produces dark pigment will tend to produce offspring that also have dark eyes, because those offspring will inherit this gene with its base-pair sequence unaltered.

23 M. Lynch, "Mutation and Human Exceptionalism: Our Future Genetic Load."

24 Such highly conserved DNA sequences show that it is possible for a large fraction of organisms to inherit these genes intact from their parents, allowing these very important base-pair sequences to persist over many generations. Of course, natural selection must also consistently weed out mutants. The most highly conserved genes are considered to be so important for key vital functions that any mutation is almost certain to be highly deleterious or lethal. For example, the structure of hemoglobin has remained quite similar at least since the common ancestor of mammals and teleost fishes over five hundred million years ago. This indicates that mutants with an altered hemoglobin structure almost always die without leaving offspring, while enough individuals inherit an intact hemoglobin gene from their parents to sustain the population.

25 Y. Erlich and D. Zielinski, "DNA Fountain Enables a Robust and Efficient Storage Architecture."

26 See K. Sterelny, K. C. Smith, and M. Dickinson, "The Extended Replicator." However, as Peter Godfrey-Smith has argued, this stable repository of biological information

need not be DNA. In principle, some other type of molecule such as a protein could play a similar role. Paul Griffiths and Russell Gray have also argued that genes have no primacy over nongenetic factors in development; both are equally essential for the formation of a living organism and the perpetuation of biological features (P. E. Griffiths and R. D. Gray, "Developmental Systems and Evolutionary Explanation").

27 There's even a special holiday called DNA Day, dedicated to the discovery of DNA's structure.

28 A recent, widely publicized genome-wide association study based on a huge sample identified dozens of new genes for human intelligence. Yet, all these genes together explained less than 5 percent of the variation in this trait (see Suzanne Sniekers et al., "Genome-Wide Association Meta-Analysis of 78,308 Individuals Identifies New Loci and Genes Influencing Human Intelligence").

29 E. Jablonka and G. Raz, "Transgenerational Epigenetic Inheritance: Prevalence, Mechanisms, and Implications for the Study of Heredity and Evolution." The qualifier "transgenerational" is added by some authors to specify the transmission of epigenetic factors through the germ line.

30 J. Beisson, "Preformed Cell Structure and Cell Heredity"; M. Bornens, "Organelle Positionining and Cell Polarity"; J. Shorter and S. Lindquist, "Prions as Adaptive Conduits of Memory and Inheritance."

31 A. J. Crean, M. I. Adler, and R. Bonduriansky, "Seminal Fluid and Mate Choice: New Predictions."

32 For example, cultural transmission can occur only in animals with brains, effects mediated by the contents of milk are only possible in mammals, and paternal effects can only occur in sexually reproducing species. Likewise, it is possible that epigenetic inheritance is more prevalent in plants, which produce gametes from somatic tissue and possess an RNA-directed mechanism for de novo DNA methylation (see M. Robertson and C. Richards, "Non-genetic Inheritance in Evolutionary Theory—the Importance of Plant Studies"), than in animals such as vertebrates and arthropods, whose germ lines are "sequestered" during embryonic development (see chapter 9 for a discussion of variation in the nature of heredity).

33 Even medical researchers are coming to realize that very different mechanisms, such as behavioral/cultural and epigenetic inheritance, can have similar transgenerational effects (see M. Pembrey et al., "Human Transgenerational Responses to Early-Life Experience: Potential Impact on Development, Health, and Biomedical Research").

34 Eva Jablonka and Marion Lamb have subdivided heredity into four dimensions (genetic, epigenetic, behavioral, and symbolic: E. Jablonka and M. J. Lamb, *Evolution in Four Dimensions*), but it's possible to imagine many other ways to slice up the pie.

35 Such effects complicate quantitative genetic studies aimed at measuring genetic effects. Whether inheritance is genetic or nongenetic, the outcome can be a positive correlation between relatives for the trait of interest. Indeed, if all we knew was the body size of parents and their offspring, we would be hard pressed to guess whether inheritance was genetic or nongenetic.

36 T. Uller, "Developmental Plasticity and the Evolution of Parental Effects."

37 Nongenetic inheritance is only one area of biology where a broader role for environment in evolution is being explored. A number of authors have argued that environmental effects on gene expression and development play a central role in evolution by

generating novel phenotypes that natural selection can act on (see M. J. West-Eberhard, *Developmental Plasticity and Evolution*; S. E. Sultan, *Organism and Environment*), and that individual organisms and populations in turn modify their environments and thereby alter patterns of natural selection on themselves and their descendants (see F. J. Odling-Smee, *Niche Construction*; A. P. Hendry, *Eco-Evolutionary Dynamics*).

38 Pembrey et al., "Human Transgenerational Responses to Early-Life Experience: Potential Impact on Development, Health, and Biomedical Research."

39 V. S. Knopik et al., "The Epigenetics of Maternal Cigarette Smoking during Pregnancy and Effects on Child Development."

40 For example, a dominant mutation of the *LMNA* gene in the germ line of one parent causes children that inherit this mutation to exhibit progeria, a condition characterized by extremely rapid aging (see L. B. Gordon et al., "Progeria: A Paradigm for Translational Medicine").

41 For an engaging overview of such effects see E. Avital and E. Jablonka, *Animal Traditions: Behavioural Inheritance in Evolution.*

42 However, truly long-term stability is likely only when vertical transmission is reliably reinforced by horizontal transmission (that is, transmission between individuals of similar age, including unrelated individuals) within sizeable populations, as shown by the rapid loss of cultural elements in small, isolated populations of humans and the loss of song syntax elements in isolated populations of birds (M. A. Kline and R. Boyd, "Population Size Predicts Technological Complexity in Oceania"; R. F. Lachlan et al., "The Progressive Loss of Syntactical Structure in Bird Song along an Island Colonization Chain").

43 For a detailed discussion of self-sustaining loops that function at a cellular level, see Jablonka and Lamb, *Evolution in Four Dimensions*; Jablonka and Raz, "Transgenerational Epigenetic Inheritance: Prevalence, Mechanisms, and Implications for the Study of Heredity and Evolution."

44 The molecular mechanisms involved in paramutation are complex and diverse, with DNA methylation, RNA interference, and chromatin structure all playing a role in various systems (see V. Chandler and M. Alleman, "Paramutation: Epigenetic Instructions Passed across Generations"; M. Haring et al., "The Role of DNA Methylation, Nucleosome Occupancy, and Histone Modifications in Paramutation"; Jablonka and Raz, "Transgenerational Epigenetic Inheritance: Prevalence, Mechanisms, and Implications for the Study of Heredity and Evolution").

45 Throughout this book, we use the term *heredity* to refer to the entire set of transmitted variation and *inheritance* to refer to transmission via a particular hereditary mechanism. Thus, the standard genetic concept of heredity encompasses just one inheritance mechanism (i.e., genetic inheritance), whereas extended heredity encompasses genetic inheritance and multiple nongenetic mechanisms of inheritance.

46 P. J. Richerson and R. Boyd, "A Dual Inheritance Model of the Human Evolutionary Process I: Basic Postulates and a Simple Model."

47 Extended heredity also shares some elements with developmental systems theory, such as the idea that heredity involves the perpetuation of developmental processes that are shaped by both genetic and nongenetic factors (e.g., see Griffiths and Gray, "Developmental Systems and Evolutionary Explanation").

48 Danchin et al., "Beyond DNA: Integrating Inclusive Inheritance into an Extended Theory of Evolution"; E. Jablonka, "Information: Its Interpretation, Its Inheritance, and

Its Sharing"; E. Danchin, "Avatars of Information: Towards an Inclusive Evolutionary Synthesis."

49 J. Maynard Smith and E. Szathmáry, *The Major Transitions in Evolution*; E. Szathmáry, "The Evolution of Replicators."

50 Genetic variation is further categorized as "additive" (that is, variation reflecting effects of distinct alleles that can be detected whenever those alleles are present in an individual's genome within a suitable environment) versus "nonadditive" (that is, "dominance" and "epistatic" variation reflecting interactions of alleles within a genome). Dominance and epistatic variance is generally not heritable unless the relevant combinations of genes can be transmitted together to offspring.

51 See Danchin et al., "Beyond DNA: Integrating Inclusive Inheritance into an Extended Theory of Evolution"; Jablonka and Lamb, *Epigenetic Inheritance and Evolution*; *Evolution in Four Dimensions*; E. Danchin and R. H. Wagner, "Inclusive Heritability: Combining Genetic and Non-genetic Information to Study Animal Behavior and Culture"; N. G. Prasad et al., "Rethinking Inheritance, yet Again: Inheritomes, Contextomes and Dynamic Phenotypes"; E. Danchin et al., "Public Information: From Noisy Neighbors to Cultural Evolution."

52 For discussions of the relationship between development and heredity, see K. N. Laland et al., "Cause and Effect in Biology Revisited: Is Mayr's Proximate-Ultimate Distinction Still Useful?"; T. Uller and H. Helanterä, "Non-genetic Inheritance in Evolutionary Theory: A Primer"; S. H. Rice, "The Place of Development in Mathematical Evolutionary Theory"; A. V. Badyaev and T. Uller, "Parental Effects in Ecology and Evolution: Mechanisms, Processes, and Implications."

53 C. R. Darwin, *On the Origin of Species*.

54 W. Johannsen, "The Genotype Conception of Heredity."

55 Mayr, *The Growth of Biological Thought: Diversity, Evolution, and Inheritance*.

56 For more comprehensive overviews of the history of heredity, see P. J. Bowler, *The Mendelian Revolution: The Emergence of Hereditarian Concepts in Modern Science and Society*; S. Müller-Wille and H-J. Rheinberger, *Heredity Produced: At the Crossroads of Biology, Politics, and Culture, 1500–1870*; *A Cultural History of Heredity*; J. Sapp, *Beyond the Gene: Cytoplasmic Inheritance and the Struggle for Authority in Genetics*; *Genesis: The Evolution of Biology*.

57 See, in particular, Jablonka and Lamb, *Epigenetic Inheritance and Evolution*; *Evolution in Four Dimensions*; Sapp, *Beyond the Gene: Cytoplasmic Inheritance and the Struggle for Authority in Genetics*; *Genesis: The Evolution of Biology*.

58 A. Koestler, *The Sleepwalkers*.

59 C. López-Beltrán, "The Medical Origins of Heredity."

60 The terms *soft* and *hard* heredity were coined by Ernst Mayr (E. Mayr, "Prologue: Some Thoughts on the History of the Evolutionary Synthesis").

61 C. Zirkle, "The Early History of the Idea of the Inheritance of Acquired Characters and of Pangenesis."

62 J-B. Lamarck, *Philosophie Zoologique*.

63 C. R. Darwin, *The Descent of Man*. Darwin's fullest treatment of this hypothesis can be found in his *The Variation of Animals and Plants under Domestication*.

64 Bowler, *The Mendelian Revolution: The Emergence of Hereditarian Concepts in Modern Science and Society*.

65 Darwin, *The Variation of Animals and Plants under Domestication*, vol. 1.

66 Darwin, *The Variation of Animals and Plants under Domestication*, 1: 392.

67 A. Weismann, *The Germ-Plasm: A Theory of Heredity*.

68 F. Galton, "Hereditary Improvement."

69 F. Galton, "Hereditary Character and Talent."

70 This compatibility of heredity with plasticity (and therefore of "nature" with "nurture") was recognized in the early days of Mendelian genetics. For example, it was already expressed very clearly by the geneticist E. G. Ford in his 1931 book *Mendelism and Evolution*.

71 NCD Risk Factor Collaboration, "A Century of Trends in Adult Human Height."

72 Weismann, *The Germ-Plasm: A Theory of Heredity*.

73 See C. E. Juliano, S. Z. Swartz, and G. M. Wessel, "A Conserved Germline Multipotency Program."

74 For example, see T. H. Morgan's discussion of Weismann's influence (T. H. Morgan, *The Theory of the Gene*, 31).

75 Johannsen, "The Genotype Conception of Heredity."

76 F. Galton, "Experiments in Pangenesis, by Breeding from Rabbits of a Pure Variety, into Whose Circulation Blood Taken from Other Varieties Had Previously Been Largely Transfused."

77 A. Weismann, *Essays upon Heredity and Kindred Biological Problems*.

78 Translation from J. A. Peters, *Classic Papers in Genetics*.

79 N. Roll-Hansen, "Sources of Wilhelm Johannsen's Genotype Theory."

80 G. M. Cook, "Neo-Lamarckian Experimentalism in America: Origins and Consequences."

81 M. D. Laubichler and E. H. Davidson, "Boveri's Long Experiment: Sea Urchin Merogones and the Establishment of the Role of Nuclear Chromosomes in Development." Boveri eventually concluded that hereditary factors were located exclusively in the nucleus (Th. Boveri, "An Organism Produced Sexually without Characteristics of the Mother").

82 For example, see L. Li, P. Zheng, and J. Dean, "Maternal Control of Early Mouse Development"; F. L. Marlow, "Maternal Control of Development in Vertebrates."

83 Sapp, *Beyond the Gene: Cytoplasmic Inheritance and the Struggle for Authority in Genetics*.

84 See A. O. Vargas, "Did Paul Kammerer Discover Epigenetic Inheritance? A Modern Look at the Controversial Midwife Toad Experiments"; G. Weismann, "The Midwife Toad and Alma Mahler: Epigenetics or a Matter of Deception?"

85 Sapp, *Beyond the Gene: Cytoplasmic Inheritance and the Struggle for Authority in Genetics*.

86 Morgan, *The Theory of the Gene*, 31.

87 Morgan, *The Physical Basis of Heredity*.

88 F. B. Hanson, "Modifications in the Albino Rat Following Treatment with Alcohol Fumes and X-Rays, and the Problem of Their Inheritance."

89 J. Lederberg, "Problems in Microbial Genetics," 153.

90 For a discussion of such difficulties, see P. J. Pauly, "How Did the Effects of Alcohol on Reproduction Become Scientifically Uninteresting?"

91 R. Bonduriansky, "Rethinking Heredity, Again."

92 Mayr, *The Growth of Biological Thought: Diversity, Evolution, and Inheritance*. A number of phenomena reminiscent of genetic encoding have since been reported. In

the early 1980s, Australian researcher E. J. Steele, working at the University of Toronto, concluded that acquired immunity in mice was incorporated into germ-line DNA sequences. He argued that this was possible because RNA released from somatic cells traveled to the gonads and was reverse-transcribed into germ-line DNA (see E. J. Steele, *Somatic Selection and Adaptive Evolution: On the Inheritance of Acquired Characters*; E. J. Steele, R. A. Lindley, and R. V. Blanden, *Lamarck's Signature: How Retrogenes Are Changing Darwin's Natural Selection Paradigm*). Although Steele's ideas have remained controversial, such a mechanism seems more plausible in light of the discovery of extracellular vesicles (see chapter 4). Furthermore, it's now known that environmental factors can lead to changes in the number of repetitive DNA sequences, such as ribosomal genes and telomeres (see chapter 5), and bacteria have been found to possess an acquired immunity system (CRISPR-Cas9) whereby fragments of viral and plasmid DNA are incorporated into the bacterial genome (see chapter 10).

93 Modified from Bonduriansky, "Rethinking Heredity, Again." The quotations come from the following sources: F.H.C. Crick, "The Croonian Lecture: The Genetic Code"; T. Dobzhansky, *Genetics and the Origin of Species*; *Genetics of the Evolutionary Process*; J.B.S. Haldane and J. Huxley, *Animal Biology*; Johannsen, "The Genotype Conception of Heredity"; Mayr, *The Growth of Biological Thought: Diversity, Evolution, and Inheritance*; Morgan, *The Theory of the Gene*; Weismann, *The Germ-Plasm: A Theory of Heredity*; W. Bateson, *William Bateson, F.R.S., Naturalist: His Essays and Addresses Together with a Short Account of His Life by Beatrice Bateson.*

94 The analogy was originally made by August Weismann in his *The Evolution Theory* (2: 63). It was reused by Julian Huxley in his book *Soviet Genetics and World Science.*

95 Galton, "Hereditary Improvement."

96 For a history of eugenics, see K. L. Garver and B. Garver, "Eugenics: Past, Present, and the Future."

97 Sapp, *Genesis: The Evolution of Biology.*

98 V. N. Soyfer, *Lysenko and the Tragedy of Soviet Science.*

99 Soyfer, *Lysenko.*

100 See W. deJong-Lambert, *The Cold War Politics of Genetic Research: An Introduction to the Lysenko Affair.*

101 Muller's departure from the USSR was motivated not just by his reaction to the rise of Lysenkoism but also by his failure to convince Stalin to adopt eugenic policies.

102 Huxley, *Soviet Genetics and World Science.*

103 DeJong-Lambert, *The Cold War Politics of Genetic Research: An Introduction to the Lysenko Affair.*

104 A prominent exception to this rule was H. J. Muller, who regarded soft heredity as inimical to socialist ideals because he believed that the harsh lives of early twentieth-century workers would thereby relegate their descendants to many generations of inferiority.

105 Bowler, *The Mendelian Revolution: The Emergence of Hereditarian Concepts in Modern Science and Society*; Sapp, *Beyond the Gene: Cytoplasmic Inheritance and the Struggle for Authority in Genetics*; *Genesis: The Evolution of Biology.*

106 The term *Modern Synthesis* was introduced by Julian Huxley in his influential 1942 book *Evolution: The Modern Synthesis*. Modern Synthesis theory is sometimes also called *neo-Darwinism.*

107 L. C. Dunn and T. Dobzhansky, *Heredity, Race, and Society*. For the most detailed exposition of this view of the history of heredity, see Mayr, *The Growth of Biological Thought: Diversity, Evolution, and Inheritance*.

108 Historian Jan Sapp has pointed out that scientists often simplify the history of their discipline into a heroic story of progress from ignorance to knowledge, in which every step is based on facts and logic, and the neglected works of visionaries are rediscovered and vindicated. The reality is often far messier (see J. Sapp, *Where the Truth Lies: Franz Moewus and the Origins of Molecular Biology*). (Of course, our own interpretation of the history of heredity is entirely accurate and free of bias.)

109 For a discussion of the changing terminology, see C. L. Richards, O. Bossdorf, and M. Pigliucci, "What Role Does Heritable Epigenetic Variation Play in Phenotypic Evolution?"

110 The narrow-sense definition of "epigenetics" is the one that readers may have encountered in books such as Nessa Carey's *The Epigenetics Revolution: How Modern Biology Is Rewriting Our Understanding of Genetics, Disease, and Inheritance*, and Richard Frances's *Epigenetics: How Environment Shapes Our Genes*. However, these books outline examples of the role of epigenetic factors both within and across generations.

111 E. J. Richards, "Inherited Epigenetic Variation—Revisiting Soft Inheritance."

112 N. A. Youngson and E. Whitelaw, "Transgenerational Epigenetic Effects."

113 Other interesting examples can be found in recent reviews, such as Y. Wang, H. Liu, and Z. Sun, "Lamarck Rises from His Grave: Parental Environment-Induced Epigenetic Inheritance in Model Organisms and Humans."

114 C. Linnaeus, *Systema Naturae*, I.

115 P. Cubas, C. Vincent, and E. Coen, "An Epigenetic Mutation Responsible for Natural Variation in Floral Symmetry."

116 David Haig has pointed out that this epimutation is much less stable than a typical genetic mutation, and the original population of *Linaria* epimutants Linnaeus studied is therefore unlikely to have survived until the present day (see D. Haig, "Weismann Rules! OK? Epigenetics and the Lamarckian Temptation").

117 Retrotransposons and other transposons (transposable elements) appear to play an important role in the evolution of epigenetic inheritance in mammals and other taxa because DNA methylation is a mechanism recruited to shut down transposon activity. The insertion of a transposon at a new location in the genome therefore results in methylation of that genomic region, and the silencing effects can leak across to adjacent genes. Variable methylation of the transposon (resulting, for example, from variation in the availability of dietary methyl donors) can therefore result in environmentally mediated epigenetic control of gene expression (as in the example of the *Agouti* gene). Transposons constitute a substantial fraction of the genomes of many organisms, and are thought to play an important role in the evolution of new genes.

118 G. L. Wolff, "Influence of Maternal Phenotype on Metabolic Differentiation of *Agouti* Locus in the Mouse"; H. D. Morgan et al., "Epigenetic Inheritance at the *Agouti* Locus in the Mouse."

119 J. E. Cropley et al., "The Penetrance of an Epigenetic Trait in Mice Is Progressively Yet Reversibly Increased by Selection and Environment."

120 J. E. Cropley et al., "Germ-Line Epigenetic Modification of the Murine A^{vy} Allele by Nutritional Supplementation."

121 M. E. Blewitt et al., "Dynamic Reprogramming of DNA Methylation at an Epigenetically Sensitive Allele in Mice."

122 M. D. Anway et al., "Epigenetic Transgenerational Actions of Endocrine Disruptors and Male Fertility."

123 A. Schuster, M. K. Skinner, and W. Yan, "Ancestral Vinclozolin Exposure Alters the Epigenetic Transgenerational Inheritance of Sperm Small Noncoding RNAs."

124 See M. Manikkam et al., "Plastics Derived Endocrine Disruptors (BPA, DEHP, and DBP) Induce Epigenetic Transgenerational Inheritance of Obesity, Reproductive Disease, and Sperm Epimutations"; E. E. Nilsson and M. K. Skinner, "Environmentally Induced Epigenetic Transgenerational Inheritance of Disease Susceptibility."

125 K. C. Calhoun et al., "Bisphenol A Exposure Alters Developmental Gene Expression in the Fetal Rhesus Macaque Uterus"; J. D. Elsworth et al., "Low Circulating Levels of Bisphenol-A Induce Cognitive Deficits and Loss of Asymmetric Spine Synapses in Dorsolateral Prefrontal Cortex and Hippocampus of Adult Male Monkeys"; A. P. Tharpa et al., "Bisphenol A Alters the Development of the Rhesus Monkey Mammary Gland"; A. Nakagami et al., "Alterations in Male Infant Behaviors towards Its Mother by Prenatal Exposure to Bisphenol A in Cynomolgus Monkeys (*Macaca fascicularis*) during Early Suckling Period."

126 P. Alonso-Magdalena, F. J. Rivera, and C. Guerrero-Bosagna, "Bisphenol-A and Metabolic Diseases: Epigenetic, Developmental, and Transgenerational Basis."

127 R. K. Bhandari, F. S. vom Saal, and D. E. Tillitt, "Transgenerational Effects from Early Developmental Exposures to Bisphenol A or 17a-Ethinylestradiol in Medaka, *Oryzias latipes*"; A. Ziv-Gal et al., "The Effects of In Utero Bisphenol A Exposure on Reproductive Capacity in Several Generations of Mice."

128 B. G. Dias and K. J. Ressler, "Parental Olfactory Experience Influences Behaviour and Neural Structure in Subsequent Generations."

129 J. A. Hackett, J. J. Zylicz, and A. Surani, "Parallel Mechanisms of Epigenetic Reprogramming in the Germline."

130 H. D. Morgan et al., "Epigenetic Reprogramming in Mammals."

131 L. Jiang et al., "Sperm, but Not Oocyte, DNA Methylome Is Inherited by Zebrafish Early Embryos"; M. E. Potok et al., "Reprogramming the Maternal Zebrafish Genome after Fertilization to Match the Paternal Methylation Pattern."

132 D. K. Seymour and C. Becker, "The Causes and Consequences of DNA Methylome Variation in Plants."

133 Q. Chen, W. Yan, and E. Duan, "Epigenetic Inheritance of Acquired Traits through Sperm RNAs and Sperm RNA Modifications."

134 L. Houri-Zeevi and O. Rechavi, "A Matter of Time: Small RNAs Regulate the Duration of Epigenetic Inheritance."

135 M. Yan et al., "A High-Throughput Quantitative Approach Reveals More Small RNA Modifications in Mouse Liver and Their Correlation with Diabetes"; K. R. Chi, "The RNA Code Comes into Focus."

136 Chen, Yan, and Duan, "Epigenetic Inheritance of Acquired Traits through Sperm RNAs and Sperm RNA Modifications."

137 L. Vojtech et al., "Exosomes in Human Semen Carry a Distinctive Repertoire of Small Non-coding RNAs with Potential Regulatory Functions."

138 I. Melentijevic et al., "*C. elegans* Neurons Jettison Protein Aggregates and Mitochondria under Neurotoxic Stress"; S. Devanapally, S. Ravikumar, and A. M. Jose,

"Double-Stranded RNA Made in *C. elegans* Neurons Can Enter the Germline and Cause Transgenerational Gene Silencing."

139 S. A. Eaton et al., "Roll over Weismann: Extracellular Vesicles in the Transgenerational Transmission of Environmental Effects."

140 M. Rassoulzadegan et al., "RNA-Mediated Non-Mendelian Inheritance of an Epigenetic Change in the Mouse."

141 K. D. Wagner et al., "RNA Induction and Inheritance of Epigenetic Cardiac Hypertrophy in the Mouse."

142 K. Gapp et al., "Implication of Sperm RNAs in Transgenerational Inheritance of the Effects of Early Trauma in Mice"; A. B. Rodgers et al., "Transgenerational Epigenetic Programming via Sperm microRNA Recapitulates Effects of Paternal Stress."

143 Q. Chen et al., "Sperm tsRNAs Contribute to Intergenerational Inheritance of an Acquired Metabolic Disorder"; V. Grandjean et al., "RNA-Mediated Paternal Heredity of Diet-Induced Obesity and Metabolic Disorders."

144 Vojtech et al., "Exosomes in Human Semen Carry a Distinctive Repertoire of Small Non-coding RNAs with Potential Regulatory Functions."

145 L. Morgado et al., "Small RNAs Reflect Grandparental Environments in Apomictic Dandelion."

146 Small RNAs appear to play a role in the methylation of histones and may form part of the machinery that allows chromatin structure to respond to environment (see Houri-Zeevi and Rechavi, "A Matter of Time: Small RNAs Regulate the Duration of Epigenetic Inheritance").

147 L. Daxinger and E. Whitelaw, "Understanding Transgenerational Epigenetic Inheritance via the Gametes in Mammals"; R. Fraser and C-J. Lin, "Epigenetic Reprogramming of the Zygote in Mice and Men: On Your Marks, Get Set, Go!"

148 V. Sollars et al., "Evidence for an Epigenetic Mechanism by Which Hsp90 Acts as a Capacitor for Morphological Evolution."

149 A. Ost et al., "Paternal Diet Defines Offspring Chromatin State and Intergenerational Obesity."

150 N. L. Vastenhouw et al., "Gene Expression: Long-Term Gene Silencing by RNAi."

151 E. L. Greer et al., "Members of the H3K4 Trimethylation Complex Regulate Lifespan in a Germline-Dependent Manner in *C. elegans*"; E. L. Greer et al., "Transgenerational Epigenetic Inheritance of Longevity in *Caenorhabditis elegans*."

152 A. Klosin et al., "Transgenerational Transmission of Environmental Information in *C. elegans*."

153 Although see, for example, Badyaev and Uller, "Parental Effects in Ecology and Evolution: Mechanisms, Processes, and Implications"; M. Kirkpatrick and R. Lande, "The Evolution of Maternal Characters."

154 Many additional examples can be found in recent books and papers. For reviews on maternal effects, see D. Maestripieri and J. M. Mateo, *Maternal Effects in Mammals*; T. A. Mousseau and C. W. Fox, *Maternal Effects as Adaptations*; Badyaev and Uller, "Parental Effects in Ecology and Evolution: Mechanisms, Processes, and Implications." For reviews on paternal effects, see A. J. Crean and R. Bonduriansky, "What Is a Paternal Effect?"; J. P. Curley, R. Mashoodh, and F. A. Champagne, "Epigenetics and the Origins of Paternal Effects"; A. Soubry et al., "A Paternal Environmental Legacy: Evidence for Epigenetic Inheritance through the Male Germ Line."

155 J. B. Wolf and M. J. Wade, "What Are Maternal Effects (and What Are They Not)?"; Badyaev and Uller, "Parental Effects in Ecology and Evolution: Mechanisms, Processes, and Implications."

156 Badyaev and Uller, "Parental Effects in Ecology and Evolution: Mechanisms, Processes, and Implications"; D. J. Marshall and T. Uller, "When Is a Maternal Effect Adaptive?"

157 In evolutionary ecology, interest in maternal effects was spurred in part by the work of Tim Mousseau, Charles Fox, and others on seed beetles during the 1990s (e.g., see T. A. Mousseau and H. Dingle, "Maternal Effects in Insect Life Histories"; T. A. Mousseau and C. W. Fox, "The Adaptive Significance of Maternal Effects"; *Maternal Effects as Adaptations*).

158 The presumed rarity of paternal effects is probably one reason why the "paternal half-sib" design became popular in quantitative genetics. While studies based on full siblings or maternal half-sibs could be compromised by common environment effects and maternal effects, it is often assumed that analysis of paternal half-sibs should yield accurate estimates of genetic (co)variance and heritability.

159 Crean and Bonduriansky, "What Is a Paternal Effect?"

160 For an introduction to indirect genetic effects, see J. B. Wolf et al., "Evolutionary Consequences of Indirect Genetic Effects."

161 V. R. Nelson, S. H. Spiezio, and J. H. Nadeau, "Transgenerational Genetic Effects of the Paternal Y Chromosome on Daughters' Phenotypes."

162 Marshall and Uller, "When Is a Maternal Effect Adaptive?"

163 A. A. Agrawal, C. Laforsch, and R. Tollrian, "Transgenerational Induction of Defences in Animals and Plants"; L. M. Holeski, G. Jander, and A. A. Agrawal, "Transgenerational Defense Induction and Epigenetic Inheritance in Plants"; R. Tolrian, "Predator-Induced Morphological Defences: Costs, Life History Shifts, and Maternal Effects in *Daphnia pulex*."

164 D. E. Dussourd et al., "Biparental Defensive Endowment of Eggs with Acquired Plant Alkaloid in the Moth *Utetheisa ornatrix*"; S. R. Smedley and T. Eisener, "Sodium: A Male Moth's Gift to Its Offspring."

165 U. R. Ernst et al., "Epigenetics and Locust Life Phase Transitions"; G. A. Miller et al., "Swarm Formation in the Desert Locust *Schistocerca gregaria*: Isolation and NMR Analysis of the Primary Maternal Gregarizing Agent"; S. R. Ott and S. M. Rogers, "Gregarious Desert Locusts Have Substantially Larger Brains with Altered Proportions Compared with the Solitarious Phase"; S. J. Simpson and G. A. Miller, "Maternal Effects on Phase Characteristics in the Desert Locust, *Schistocerca gregaria*: A Review of Current Understanding"; S. Tanaka and K. Maeno, "A Review of Maternal and Embryonic Control of Phase-Dependent Progeny Characteristics in the Desert Locust."

166 In fact, evidence that anticipatory effects typically increase fitness is equivocal (see T. Uller, S. Nakagawa, and S. English, "Weak Evidence for Anticipatory Parental Effects in Plants and Animals").

167 See O. Leimar and J. M. McNamara. "The Evolution of Transgenerational Integration of Information in Heterogeneous Environments."

168 K. E. McGhee and A. M. Bell, "Paternal Care in a Fish: Epigenetics and Fitness Enhancing Effects on Offspring Anxiety"; K. E. McGhee et al., "Maternal Exposure to Predation Risk Decreases Offspring Antipredator Behaviour and Survival in Three-spined Stickleback."

169 E. M. Hollams et al., "Persistent Effects of Maternal Smoking during Pregnancy on Lung Function and Asthma in Adolescents"; Knopik et al., "The Epigenetics of Maternal Cigarette Smoking during Pregnancy and Effects on Child Development"; F. M. Leslie, "Multigenerational Epigenetic Effects of Nicotine on Lung Function"; S. Moylan et al., "The Impact of Maternal Smoking during Pregnancy on Depressive and Anxiety Behaviors in Children: The Norwegian Mother and Child Cohort Study."

170 For example, see B. A. Carnes, R. Riesch, and I. Schlupp, "The Delayed Impact of Parental Age on Offspring Mortality in Mice"; K. E. Gribble et al., "Maternal Caloric Restriction Partially Rescues the Deleterious Effects of Advanced Maternal Age on Offspring"; M. J. Hercus and A. A. Hoffmann, "Maternal and Grandmaternal Age Influence Offspring Fitness in *Drosophila*"; S. Kern et al., "Decline in Offspring Viability as a Manifestation of Aging in *Drosophila melanogaster*"; R. Torres, H. Drummond, and A. Velando, "Parental Age and Lifespan Influence Offspring Recruitment: A Long-Term Study in a Seabird."

171 D. J. Marshall and T. Uller, "When Is a Maternal Effect Adaptive?"; T. Uller and I. Pen, "A Theoretical Model of the Evolution of Maternal Effects under Parent-Offspring Conflict"; B. Kuijper and R. A. Johnstone, "Maternal Effects and Parent-Offspring Conflict."

172 More precisely, natural selection favors strategies that maximize "inclusive fitness," which reflects the fitness of the focal individual plus that of its relatives, with each relative's contribution to the focal individual's fitness weighted by its degree of relatedness (that is, its probability of sharing genetic alleles with the focal individual). This is why the brilliant and eccentric biologist J.B.S. Haldane once quipped that he would sacrifice his life for two brothers (each of whom shares 50 percent of his alleles, on average) or eight first cousins (each of whom shares 12.5 percent of his alleles, on average).

173 For the classic analysis of the problem of optimally trading off offspring quality against offspring number, see C. C. Smith and S. D. Fretwell, "The Optimal Balance between Size and Number of Offspring."

174 Haig invoked genomic imprinting as the mechanism that males use to obtain a greater share of maternal resources for their offspring (see D. Haig, "The Kinship Theory of Genomic Imprinting"). Genomic imprinting involves differential methylation of alleles in eggs versus sperm, resulting in epigenetic marks that persist in the embryo and can differentially regulate expression of maternally and paternally inherited alleles. For example, expression of the *Igf2* gene in the mammalian placenta regulates embryonic growth rate, and this gene is imprinted such that only the paternally inherited allele is expressed while the maternally inherited allele is silent. Haig suggested that imprinting of *Igf2* originally evolved as a paternal strategy enabling offspring to extract a greater share of maternal resources and grow faster, with subsequent evolution of a maternal counter-strategy of shutting down expression of this gene, ultimately reaching a stalemate. However, males could evolve other mechanisms, such as paternal effects mediated by factors in the seminal fluid, to obtain a greater share of maternal resources for their offspring (R. Bonduriansky, "The Ecology of Sexual Conflict: Background Mortality Can Modulate the Effects of Male Manipulation on Female Fitness").

175 R. Bonduriansky and M. Head, "Maternal and Paternal Condition Effects on Offspring Phenotype in *Telostylinus angusticollis* (Diptera : Neriidae)"; A. J. Crean, A. M. Kopps, and R. Bonduriansky, "Revisiting Telegony: Offspring Inherit an Acquired Characteristic of Their Mother's Previous Mate."

176 Effects of paternal larval diet on offspring phenotype have been reported in *Drosophila melanogaster* as well: see T. M. Valtonen et al., "Transgenerational Effects of Parental Larval Diet on Offspring Development Time, Adult Body Size, and Pathogen Resistance in *Drosophila melanogaster*"; R. K. Vijendravarma, S. Narasimha, and T. J. Kawecki, "Effects of Parental Larval Diet on Egg Size and Offspring Traits in *Drosophila*."

177 Crean, Kopps, and Bonduriansky, "Revisiting Telegony: Offspring Inherit an Acquired Characteristic of Their Mother's Previous Mate."

178 A telegony-like effect has now also been reported in *Drosophila* (F. Garcia-Gonzalez and D. K. Dowling, "Transgenerational Effects of Sexual Interactions and Sexual Conflict: Non-Sires Boost the Fecundity of Females in the Following Generation").

179 For example, see J. J. Cowley and R. D. Griesel, "The Effect on Growth and Behaviour of Rehabilitating First and Second Generation Low Protein Rats"; S. Zamenhof, E. van Marthens, and L. Grauel, "DNA (Cell Number) in Neonatal Brain: Second Generation (F2) Alternation by Maternal (F0) Dietary Protein Restriction"; S. Zamenhof, E. van Marthens, and F. L. Margolis, "DNA (Cell Number) and Protein in Neonatal Brain: Alternation by Maternal Dietary Protein Restriction."

180 Kamimae-Lanning et al., "Maternal High-Fat Diet and Obesity Compromise Fetal Hematopoiesis"; V. Amarger et al., "Protein Content and Methyl Donors in Maternal Diet Interact to Influence the Proliferation Rate and Cell Fate of Neural Stem Cells in Rat Hippocampus."

181 S-F. Ng et al., "Chronic High-Fat Diet in Fathers Programs B-Cell Dysfunction in Female Rat Offspring."

182 C. Schmauss, Z. Lee-McDermott, and L. R. Medina, "Trans-Generational Effects of Early Life Stress: The Role of Maternal Behaviour."

183 F. Pittet et al., "Effects of Maternal Experience on Fearfulness and Maternal Behaviour in a Precocial Bird."

184 M. F. Neuwald et al., "Transgenerational Effects of Maternal Care Interact with Fetal Growth and Influence Attention Skills at 18 Months of Age."

185 R. Mashoodh et al., "Paternal Social Enrichment Effects on Maternal Behavior and Offspring Growth."

186 C.R.M. Frazier et al., "Paternal Behavior Influences Development of Aggression and Vasopressin Expression in Male California Mouse Offspring."

187 McGhee and Bell, "Paternal Care in a Fish: Epigenetics and Fitness Enhancing Effects on Offspring Anxiety."

188 Mesoudi, *Cultural Evolution: How Darwinian Theory Can Explain Human Culture and Synthesize the Social Sciences.*

189 K. Sterelny, *The Evolved Apprentice.*

190 L. V. Luncz, R. Mundry, and C. Boesch, "Evidence for Cultural Differences between Neighboring Chimpanzee Communities"; A. Whiten et al., "Charting Cultural Variation in Chimpanzees."

191 See F. de Waal, *Are We Smart Enough to Know How Smart Animals Are?*

192 E.J.C. Van Leeuwen, K. A. Cronin, and D.B.M. Haun, "A Group-Specific Arbitrary Tradition in Chimpanzees (*Pan troglodytes*)."

193 M. Kawai, "Newly-Acquired Pre-Cultural Behaviour of the Natural Troop of Japanese Monkeys on Koshima Islet"; T. Matsuzawa and W. C. McGraw, "Kinji Imanishi and 60 Years of Japanese Primatology."

194 M. Krutzen et al., "Cultural Transmission of Tool Use in Bottlenose Dolphins."

195 D.W.A. Noble, R. W. Byrne, and M. J. Whiting, "Age-Dependent Social Learning in a Lizard."

196 H. Slabbekoorn and T. B. Smith, "Bird Song, Ecology, and Speciation."

197 G. M. Kozak, M. L. Head, and J. W. Boughman, "Sexual Imprinting on Ecologically Divergent Traits Leads to Sexual Isolation in Sticklebacks."

198 R. F. Lachlan and M. R. Servedio, "Song Learning Accelerates Allopatric Speciation."

199 L. M. Aplin et al., "Experimentally Induced Innovations Lead to Persistent Culture via Conformity in Wild Birds."

200 L. Boto, "Horizontal Gene Transfer in Evolution: Facts and Challenges"; S. M. Soucy, J. Huang, and J. P. Gogarten, "Horizontal Gene Transfer: Building the Web of Life."

201 See S. F. Gilbert, J. Sapp, and A. I. Tauber, "A Symbiotic View of Life: We Have Never Been Individuals"; N. A. Moran and D. B. Sloan, "The Hologenome Concept: Helpful or Hollow?"

202 G. Sharon et al., "Commensal Bacteria Play a Role in Mating Preference of *Drosophila melanogaster*"; A. Vilcinskas et al., "Invasive Harlequin Ladybird Carries Biological Weapons against Native Competitors"; J. H. Werren, L. Baldo, and M. E. Clark, "*Wolbachia*: Master Manipulators of Invertebrate Biology."

203 N. G. Rossen et al., "Fecal Microbiota Transplantation as Novel Therapy in Gastroenterology: A Systematic Review."

204 M. F. Camus, D. J. Clancy, and D. K. Dowling, "Mitochondria, Maternal Inheritance, and Male Aging"; M. F. Camus et al., "Single Nucleotides in the mtDNA Sequence Modify Mitochondrial Molecular Function and Are Associated with Sex-Specific Effects on Fertility and Aging"; G. Arnqvist et al., "Genetic Architecture of Metabolic Rate: Environment Specific Epistasis between Mitochondrial and Nuclear Genes in an Insect"; H. Lovelie et al., "The Influence of Mitonuclear Genetic Variation on Personality in Seed Beetles."

205 H. Makino et al., "Mother-to-Infant Transmission of Intestinal Bifidobacterial Strains Has an Impact on the Early Development of Vaginally Delivered Infant's Microbiota"; M. Nieuwdorp et al., "Role of the Microbiome in Energy Regulation and Metabolism."

206 M. S. LaTuga, A. Stuebe, and P. C. Seed, "A Review of the Source and Function of Microbiota in Breast Milk."

207 A. B. Javurek et al., "Discovery of a Novel Seminal Fluid Microbiome and Influence of Estrogen Receptor Alpha Genetic Status"; Corrigendum; J. White et al., "Sexually Transmitted Bacteria Affect Female Cloacal Assemblages in a Wild Bird."

208 See Nanney, "Cortical Patterns in Cellular Morphogenesis," for an excellent overview of this work.

209 H. S. Jennings, "Formation, Inheritance, and Variation of the Teeth in *Difflugia corona*: A Study of the Morphogenic Activities of Rhizopod Protoplasm."

210 J. Beisson and T. M. Sonneborn, "Cytoplasmic Inheritance of the Organization of the Cell Cortex in *Paramecium aurelia*."

211 Beisson, "Preformed Cell Structure and Cell Heredity."

212 Y. Shirokawa and M. Shimada, "Cytoplasmic Inheritance of Parent-Offspring Cell Structure in the Clonal Diatom *Cyclotella meneghiniana*."

213 F. F. Moreira-Leite et al., "A Trypanosome Structure Involved in Transmitting Cytoplasmic Information during Cell Division."

214 G. W. Grimes, "Pattern Determination in Hypotrich Ciliates."

215 For example, see S. Vaughan and H. R. Dawe, "Common Themes in Centriole and Centrosome Movements."

216 Beisson, "Preformed Cell Structure and Cell Heredity"; Nanney, "Cortical Patterns in Cellular Morphogenesis"; T. M. Sonneborn, "Does Preformed Cell Structure Play an Essential Role in Cell Heredity?"

217 Sapp, *Genesis: The Evolution of Biology*, 209. This book provides a very lucid overview of the debate on the role of preexisting structure in development.

218 C. Sardet et al., "Structure and Function of the Egg Cortex from Oogenesis through Fertilization."

219 See Li, Zheng, and Dean, "Maternal Control of Early Mouse Development"; Marlow, "Maternal Control of Development in Vertebrates."

220 K. Piotrowska and M. Zernicka-Goetz, "Role for Sperm in Spatial Patterning of the Early Mouse Embryo."

221 Bornens, "Organelle Positionining and Cell Polarity."

222 For example, the tight link between cuticular and cytoskeletal structure is suggested by work on the flagellated single-celled parasite *Trypanosoma brucei* (see S. Lacomble et al., "Basal Body Movements Orchestrate Membrane Organelle Division and Cell Morphogenesis in *Trypanosoma brucei*"; S. Y. Sun et al., "An Intracellular Membrane Junction Consisting of Flagellum Adhesion Glycoproteins Links Flagellum Biogenesis to Cell Morphogenesis in *Trypanosoma brucei*"; Moreira-Leite et al., "A Trypanosome Structure Involved in Transmitting Cytoplasmic Information during Cell Division").

223 T. Cavalier-Smith, "The Membranome and Membrane Heredity in Development and Evolution."

224 R. Halfmann and S. Lindquist, "Epigenetics in the Extreme: Prions and the Inheritance of Environmentally Acquired Traits"; Shorter and Lindquist, "Prions as Adaptive Conduits of Memory and Inheritance."

225 J. Bremer et al., "Axonal Prion Protein Is Required for Peripheral Myelin Maintenance"; L. Fioriti et al., "The Persistence of Hippocampal-Based Memory Requires Protein Synthesis Mediated by the Prion-Like Protein CPEB3."

226 R. Halfmann, S. Alberti, and S. Lindquist, "Prions, Protein Homeostasis, and Phenotypic Diversity."

227 For example, structural changes in the mitochondria of yeast are transmitted by an unknown nongenetic mechanism (see D. Lockshon, "A Heritable Structural Alteration of the Yeast Mitochondrion"). Many other examples of non-Mendelian inheritance of structure in fungi could also involve prions or other, as yet unknown, cytoplasmic elements (for example, see L. Benkemoun and S. J. Saupe, "Prion Proteins as Genetic Material in Fungi"; F. Malagnac and P. Silar, "Non-Mendelian Determinants of Morphology in Fungi").

228 Some basic experimental designs are described in R. Bonduriansky, A. J. Crean, and T. Day, "The Implications of Nongenetic Inheritance for Evolution in Changing Environments."

229 See M. K. Skinner, "Environmental Epigenetics and a Unified Theory of the Molecular Aspects of Evolution: A Neo-Lamarckian Concept That Facilitates Neo-Darwinian Evolution."

230 For details of this approach and some of the insights that it has provided, see S. J. Simpson and D. Raubenheimer, *The Nature of Nutrition: A Unifying Framework from Animal Adaptation to Human Obesity.*

231 R. Bonduriansky, A. Runagall-McNaull, and A. J. Crean, "The Nutritional Geometry of Parental Effects: Maternal and Paternal Macronutrient Consumption and Offspring Phenotype in a Neriid Fly."

232 For a discussion of the complexity of this problem, see M. Szyf, "Lamarck Revisited: Epigenetic Inheritance of Ancestral Odor Fear Conditioning."

233 A. Sharma, "Transgenerational Epigenetic Inheritance: Focus on Soma to Germline Information Transfer"; Robertson and Richards, "Non-genetic Inheritance in Evolutionary Theory—the Importance of Plant Studies."

234 A further complication is that the nature of the nongenetic factor transmitted from parent to offspring can change from one generation to another. For example, a study on mice suggested that the effects of stress were transmitted from affected males to their offspring via sperm-borne noncoding RNA. However, although the male offspring transmitted these symptoms to their own offspring, transmission to the grand-offspring did not occur via sperm-borne RNA and must therefore have involved some other nongenetic factor (Gapp et al., "Implication of Sperm RNAs in Transgenerational Inheritance of the Effects of Early Trauma in Mice").

235 For example, see S. Liu et al., "Natural Epigenetic Variation in Bats and Its Role in Evolution"; E. V. Avramidou et al., "Beyond Population Genetics: Natural Epigenetic Variation in Wild Cherry (*Prunus avium*)"; S. Hirsch, R. Baumberger, and U. Grossniklaus, "Epigenetic Variation, Inheritance, and Selection in Plant Populations"; C. M. Herrera and P. Bazaga, "Untangling Individual Variation in Natural Populations: Ecological, Genetic, and Epigenetic Correlates of Long-Term Inequality in Herbivory"; C. L. Richards, O. Bossdorf, and K.J.F. Verhoeven, "Understanding Natural Epigenetic Variation"; C. L. Richards et al., "Ecological Plant Epigenetics: Evidence from Model and Non-model Species, and the Way Forward."

236 M. J. Dubin et al., "DNA Methylation in *Arabidopsis* Has a Genetic Basis and Shows Evidence of Local Adaptation."

237 Richards et al., "Ecological Plant Epigenetics: Evidence from Model and Non-model Species, and the Way Forward"; Richards, Bossdorf, and Pigliucci, "What Role Does Heritable Epigenetic Variation Play in Phenotypic Evolution?"

238 For example, see Dominguez-Salas et al., "Maternal Nutrition at Conception Modulates DNA Methylation of Human Metastable Epialleles"; Waterland et al., "Season of Conception in Rural Gambia Affects DNA Methylation at Putative Human Metastable Epialleles."

239 Although the molecular mechanism involved is not known, the authors hypothesized that protein consumption ramps up the transcription of ribosomal genes, leading to the formation of loops of repetitive ribosomal DNA that are then prone to detaching from the chromosome (J. C. Aldrich and K. A. Maggert, "Transgenerational Inheritance of Diet-Induced Genome Rearrangements in *Drosophila*").

240 D.T.A. Eisenberg, "Inconsistent Inheritance of Telomere Length (TL): Is Offspring TL More Strongly Correlated with Maternal or Paternal TL?"; E. S. Epel et al., "Accelerated Telomere Shortening in Response to Life Stress."

241 G. R. Price, "The Nature of Selection." This paper was written circa 1971, and published posthumously.

242 Claude Shannon was an American mathematician and engineer who is credited with inventing the field of information theory. Interestingly, despite being best known for these results from engineering, his PhD thesis actually developed a mathematical framework for modeling Mendelian genetic inheritance. His thesis is freely available at http://dspace.mit.edu/handle/1721.1/11174.

243 G. R. Price, "Science and the Supernatural"; "Where Is the Definitive Experiment?"

244 O. Harman, *The Price of Altruism*.

245 G. R. Price, "Selection and Covariance."

246 G. R. Price, "The Nature of Selection."

247 The covariance between two random variables, X and Y, is defined to be the average value of X times Y, minus the average value of X times the average value of Y. In other words, using overbars for averages, we have $\mathrm{cov}(X,Y) = \overline{XY} - \overline{X}\,\overline{Y}$. In Price's equation the random variables are w and z, and because the average value of w is 1, we get $\overline{zw} - \bar{z}\overline{w} = \mathrm{cov}(z,w)$.

248 H. Helanterä and T. Uller, "The Price Equation and Extended Inheritance."

249 For an excellent treatment of Price's contributions to evolutionary biology, see S. A. Frank, "George Price's Contributions to Evolutionary Genetics"; A. Gardner, "The Price Equation."

250 S. Wright, "The Roles of Mutation, Inbreeding, Crossbreeding, and Selection in Evolution."

251 T. Day and R. Bonduriansky, "A Unified Approach to the Evolutionary Consequences of Genetic and Nongenetic Inheritance"; F. D. Klironomos, J. Berg, and S. Collins, "How Epigenetic Mutations Can Affect Genetic Evolution: Model and Mechanism"; M. Lachmann and E. Jablonka, "The Inheritance of Phenotypes: An Adaptation to Fluctuating Environments"; C. Pál and I. Miklós, "Epigenetic Inheritance, Genetic Assimilation and Speciation"; A. P. Feinberg and R. A. Irizarry, "Stochastic Epigenetic Variation as a Driving Force of Development, Evolutionary Adaptation, and Disease."

252 More generally we can include an additional copy of Price's equation for each additional genetic or nongenetic component of inheritance that is of interest (see Bonduriansky and Day, "Nongenetic Inheritance and Its Evolutionary Implications"; Day and Bonduriansky, "A Unified Approach to the Evolutionary Consequences of Genetic and Nongenetic Inheritance").

253 Richards, "Inherited Epigenetic Variation—Revisiting Soft Inheritance."

254 C. A. Hutchison III et al., "Design and Synthesis of a Minimal Bacterial Genome"; C. Lartigue et al., "Genome Transplantation in Bacteria: Changing One Species to Another"; D. G. Gibson et al., "Creation of a Bacterial Cell Controlled by a Chemically Synthesized Genome."

255 C. Lartigue et al., "Genome Transplantation in Bacteria: Changing One Species to Another."

256 These examples (and other examples that we explore in chapter 9) support the idea that genes can be "followers, not leaders" in evolution—that is, that evolution can be driven at least initially by selection and response of nongenetic traits. This view was espoused by Mary Jane West-Eberhard in her book *Developmental Plasticity and Evolution* in relation to the role of classic developmental plasticity. West-Eberhard argued that many instances of adaptation begin with the release of novel phenotypic variation in response to changed environmental conditions, followed by genetic assimilation of

advantageous phenotypes, whereby these phenotypes come to be expressed even in the absence of the original inducing environment. Jablonka and Lamb have also pointed out that genes can be followers in evolution with nongenetic inheritance.

257 Y. Itan et al., "A Worldwide Correlation of Lactase Persistence Phenotype and Genotypes"; D. M. Swallow, "Genetics of Lactase Persistence and Lactose Intolerance."

258 R. E. Green et al., "A Draft Sequence of the Neandertal Genome."

259 N. Swaminathan, "Not Milk? Neolithic Europeans Couldn't Stomach the Stuff."

260 See comments on Jablonka and Lamb's *Epigenetic Inheritance and Evolution* in the *Journal of Evolutionary Biology*, vol. 11, issue 2 (1998). For more recent critiques, see B. Dickins and Q. Rahman, "The Extended Evolutionary Synthesis and the Role of Soft Inheritance in Evolution"; T. E. Dickins and B.J.A. Dickins, "Mother Nature's Tolerant Ways: Why Non-genetic Inheritance Has Nothing to Do with Evolution"; D. J. Futuyma, "Can Modern Evolutionary Theory Explain Macroevolution?"; Haig, "Weismann Rules! OK? Epigenetics and the Lamarckian Temptation."

261 R. E. Furrow, "Epigenetic Inheritance, Epimutation, and the Response to Selection"; J. L. Geoghegan and H. G. Spencer, "Exploring Epiallele Stability in a Population-Epigenetic Model."

262 See C. L. Caprette et al., "The Origin of Snakes (Serpentes) as Seen through Eye Anatomy"; B. F. Simões et al., "Visual System Evolution and the Nature of the Ancestral Snake."

263 Klironomos, Berg, and Collins, "How Epigenetic Mutations Can Affect Genetic Evolution: Model and Mechanism."

264 Day and Bonduriansky, "A Unified Approach to the Evolutionary Consequences of Genetic and Nongenetic Inheritance."

265 Haig, "Weismann Rules! OK? Epigenetics and the Lamarckian Temptation"; Dickins and Rahman, "The Extended Evolutionary Synthesis and the Role of Soft Inheritance in Evolution"; Dickins and Dickins, "Mother Nature's Tolerant Ways: Why Non-genetic Inheritance Has Nothing to Do with Evolution"; Futuyma, "Can Modern Evolutionary Theory Explain Macroevolution?"

266 Maynard Smith and Szathmáry, *The Major Transitions in Evolution*.

267 Jablonka and Lamb, *Evolution in Four Dimensions*.

268 P. Godfrey-Smith, "Is It a Revolution?"

269 P. Godfrey-Smith, "Is It a Revolution?"

270 It is sometimes argued that DNA has greater combinatorial complexity than nongenetic mechanisms because nongenetic factors cannot take on as many potential states, but the issue is actually a bit subtler than is sometimes appreciated. For example, a single truly analogue nongenetic factor actually has a greater number of possible states than does a genome of any size. Formally, the set of possible states of a single analogue factor is uncountable, and so the size of this set (technically called the set's cardinality) is the same as the cardinality of the set of real numbers. On the other hand, the set of possible states of any genome is always countable and so it has a cardinality equal to or less than that of the set of integers. The former is, in a precise sense, infinitely larger than the latter. Of course, many of the different states of a single analogue nongenetic factor are likely to be selectively equivalent, and so the number of selectively relevant states is probably much smaller than the number of possible states. At the same time, however, this is probably also true of the states of the genome.

271 Modern Synthesis theory recognizes that environmental factors such as ionizing radiation or chemical mutagens can induce genetic mutations in the germ line, and thereby influence the phenotype of descendants. As we noted in chapter 2, the key distinction between such mutations and acquired traits transmitted by nongenetic inheritance is consistency. Genetic mutations are assumed to be unpredictable and unrepeatable, whereas nongenetic factors can change in consistent ways in response to specific environmental factors. For example, one hundred individuals exposed to ionizing radiation might all suffer different germ-line mutations. By contrast, if those same individuals are fed a sugar-rich diet, they might transmit consistent physiological effects to their offspring.

272 Quoted in Sapp, "Cytoplasmic Heretics."

273 Maternal effects are routinely incorporated into quantitative-genetic analyses, but paternal effects rarely are. More generally, although some recent efforts have been made to incorporate nongenetic factors into quantitative-genetic experimental designs and statistical models (e.g., F. Johannes and M. Colome-Tatche, "Quantitative Epigenetics through Epigenomic Perturbation of Isogenic Lines"; O. Tal, E. Kisdi, and E. Jablonka, "Epigenetic Contribution to Covariance between Relatives"; Z. Wang et al., "A Quantitative Genetic and Epigenetic Model of Complex Traits"; A. W. Santure and H. G. Spencer, "Influence of Mom and Dad: Quantitative Genetic Models for Maternal Effects and Genomic Imprinting"; Danchin et al., "Beyond DNA: Integrating Inclusive Inheritance into an Extended Theory of Evolution"; Danchin et al., "Public Informatikon: From Noisy Neighbors to Cultural Evolution"; Danchin and Wagner, "Inclusive Heritability: Combining Genetic and Non-genetic Information to Study Animal Behavior and Culture"), none of these can discriminate all possible types of nongenetic effects. Thus, quantitative-genetic analyses potentially incorporate nongenetic sources of heritable variation into estimates of genetic variance. Such nongenetic effects could be very important for some traits, and establishing their role will require experimentation and investigation of proximate mechanisms (see chapter 5).

274 This idea is supported by the observation that identical twins become less epigenetically similar over the course of their lives (M. F. Fraga et al., "Epigenetic Differences Arise during the Lifetime of Monozygotic Twins").

275 Furrow, Christiansen, and Feldman, "Environment-Sensitive Epigenetics and the Heritability of Complex Diseases"; Manolio et al., "Finding the Missing Heritability of Complex Diseases."

276 M. J. West-Eberhard, "Dancing with DNA and Flirting with the Ghost of Lamarck."

277 Laland et al., "The Extended Evolutionary Synthesis: Its Structure, Assumptions, and Predictions."

278 Jablonka and Lamb, *Evolution in Four Dimensions.*

279 See Futuyma, "Can Modern Evolutionary Theory Explain Macroevolution?"; Haig, "Weismann Rules! OK? Epigenetics and the Lamarckian Temptation."

280 A deeper issue considered by some authors is whether it is even possible to cleanly separate natural selection from the production of variation. The majority of contemporary evolutionary analysis is based on an assumption that this separation is possible, but that need not be true in real biological systems. If one instead chose to develop evolutionary theory based on an assumption that natural selection and the production

of variation are inextricably intertwined, then it is no longer even meaningful to ask if there might be directed variation on which selection acts (A. V. Badyaev, "Origin of the Fittest: Link between Emergent Variation and Evolutionary Change as a Critical Question in Evolutionary Biology"; T. Uller and H. Helanterä, "Niche Construction and Conceptual Change in Evolutionary Biology").

281 Bonduriansky and A. J. Crean, "What Are Parental Condition-Transfer Effects and How Can They Be Detected?"; Marshall and Uller, "When Is a Maternal Effect Adaptive?"; S. R. Proulx and H. Teotónio, "What Kind of Maternal Effects Can Be Selected for in Fluctuating Environments?"

282 Adaptive parental effects also present a sharp contrast with the random segregation of Mendelian alleles during the formation of haploid gametes (meiosis). If the two alleles present at a given locus in your genome differ in their effects on fitness such that your offspring would do better by inheriting one allele rather than the other, this difference will bear no relation to these alleles' probabilities of ending up in a gamete. Approximately half of your gametes will end up carrying the "good" allele and half will end up carrying the "bad" allele. Every genome is a randomly assembled mosaic of parental alleles, and it's a matter of pure luck how many "good" alleles an individual inherits from its parents. Contrast this Mendelian raffle with the pattern generated by adaptive parental effects; here, it's as if an individual possessing a good allele were somehow able to insure that the particular allele and not its inferior partner at the locus was consistently inserted into each gamete, and the parent could even choose which of its alleles was best depending on the environment that its offspring were expected to encounter.

283 Y. N. Harari, *Homo Deus: A Brief History of Tomorrow*.

284 Indeed, cultural evolution theory shows that maladaptive behaviors can readily spread through populations if their probability of transmission is high enough to overcome their selective disadvantage (for example, see Cavalli-Sforza and Feldman, *Cultural Transmission and Evolution: A Quantitative Approach*; Richerson and Boyd, *Not by Genes Alone: How Culture Transforms Human Evolution*). It's not difficult to find examples from human history that support this prediction. The same is true of niche construction, the process whereby organisms modify their environment through their own activities, thereby also changing the way natural selection acts on them. Proponents of niche construction have argued that this process tends to generate adaptive outcomes, citing examples such as the dams built by beavers (e.g., see F. J. Odling-Smee, K. N. Laland, and M. W. Feldman, *Niche Construction: The Neglected Process in Evolution*; Uller and Helanterä, "Niche Construction and Conceptual Change in Evolutionary Biology"). However, while adaptive examples can certainly be found, many examples of environmental modification are maladaptive: through their own activities, populations often overexploit their resources, foul their surroundings, or otherwise harm their descendants, thereby destroying rather than constructing their niche.

285 For an interesting example of the complex and unforeseeable long-term consequences of behavior that's clearly adaptive in the short term, see R. M. Sapolsky and L. J. Share, "A Pacific Culture among Wild Baboons: Its Emergence and Transmission."

286 Skinner, "Environmental Epigenetics and a Unified Theory of the Molecular Aspects of Evolution: A Neo-Lamarckian Concept That Facilitates Neo-Darwinian Evolution."

287 R. Watson and E. Szathmáry, "How Can Evolution Learn?"

288 F. Johannes et al., "Assessing the Impact of Transgenerational Epigenetic Variation on Complex Traits"; Richards et al., "Ecological Plant Epigenetics: Evidence from Model and Non-model Species, and the Way Forward."

289 Cropley et al., "The Penetrance of an Epigenetic Trait in Mice Is Progressively yet Reversibly Increased by Selection and Environment"; Sollars et al., "Evidence for an Epigenetic Mechanism by Which Hsp90 Acts as a Capacitor for Morphological Evolution"; Vastenhouw et al., "Gene Expression: Long-Term Gene Silencing by RNAi."

290 J. Liao et al., "Targeted Disruption of *DNMT1*, *DNMT3A*, and *DNMT3B* in Human Embryonic Stem Cells."

291 A. Vojta et al., "Repurposing the CRISPR-Cas9 System for Targeted DNA Methylation"; J. I. McDonald et al., "Reprogrammable CRISPR/Cas9-Based System for Inducing Site-Specific DNA Methylation."

292 O. O. Abuddayeh et al., "C2c2 Is a Single-Component Programmable RNA-Guided RNA-Targeting Crispr Effector."

293 For example, Cropley et al., "The Penetrance of an Epigenetic Trait in Mice Is Progressively yet Reversibly Increased by Selection and Environment"; Sollars et al., "Evidence for an Epigenetic Mechanism by Which Hsp90 Acts as a Capacitor for Morphological Evolution"; Vastenhouw et al., "Gene Expression: Long-Term Gene Silencing by RNAi."

294 For example, see Avramidou et al., "Beyond Population Genetics: Natural Epigenetic Variation in Wild Cherry (*Prunus avium*)"; S. Baldanzi et al., "Epigenetic Variation among Natural Populations of the South African Sandhopper *Talorchestia capensis*"; Herrera and Bazaga, "Untangling Individual Variation in Natural Populations: Ecological, Genetic, and Epigenetic Correlates of Long-Term Inequality in Herbivory"; Liu et al., "Natural Epigenetic Variation in Bats and Its Role in Evolution"; R. J. Schmitz et al., "Patterns of Population Epigenomic Diversity."

295 For example, see Badyaev and Uller, "Parental Effects in Ecology and Evolution: Mechanisms, Processes, and Implications"; Klironomos, Berg, and Collins, "How Epigenetic Mutations Can Affect Genetic Evolution: Model and Mechanism"; J. L. Geoghegan and H. G. Spencer, "Population-Epigenetic Models of Selection"; J. L. Geoghegan and H. G. Spencer, "The Evolutionary Potential of Paramutation: A Population-Epigenetic Model."

296 The Red Queen terminology was first introduced into the evolutionary literature in 1973 by Leigh Van Valen (1935–2010) in his paper L. Van Valen, "A New Evolutionary Law." This publication, along with other interesting information about Van Valen, can be found at www.leighvanvalen.com.

297 Harold Flor (1900–1991), a plant pathologist at North Dakota State University, first introduced the idea of the gene-for-gene mechanism.

298 J. N. Thompson and J. J. Burdon, "Gene-for-Gene Coevolution between Plants and Parasites."

299 The Great Famine, also sometimes colloquially referred to as the Irish Potato Famine, was caused by the introduction of the plant pathogen *Phytophthora infestans* from North America.

300 T. Kasuga and M. Gijzen, "Epigenetics and the Evolution of Virulence."

301 Suppose the frequency of allele D is p. If all genotypes are formed randomly, then DD individuals are formed by two independent draws of a D allele, giving a frequency of p^2. And EE individuals are formed by two independent draws of an E allele

(which has frequency $1 - p$), giving a frequency of $(1 - p)^2$. Finally, ED individuals are formed by either drawing an E and then a D allele (with probability $(1 - p)p$) or a D and then an E allele (with probability $p(1 - p)$). This gives a total probability of $2p(1 - p)$. These are referred to as the Hardy-Weinberg equilibrium frequencies. When $p = 0.5$ we get the frequencies given in the Box.

302 To apply the theory from chapter 6, we first need to define the trait z. We take this to be the frequency of the D allele *within an individual*. Therefore $z = 1$ for DD individuals, $z = 1/2$ for ED individuals, and $z = 0$ for EE individuals. Then the average value of z over all individuals in the population is the frequency of allele D in the population; that is, $E[z] = p$, where p is the population frequency of allele D. If we assume that mating is random, then the genotype frequencies will be in Hardy-Weinberg equilibrium at the start of each generation: that is, for DD:DN:EE we have the frequencies $p^2 : 2p(1 - p) : (1 - p)^2$. Taking the fitness values specified in Box 9.1, the population mean fitness is $E[W] = p^2 W + 2p(1-p)\frac{W}{2} = pW$. The covariance term in the Price equation (chapter 6) therefore becomes $\text{cov}(z, W) = E[zW] - E[z]E[W] = (p^2 \times 1 \times \frac{W}{E[W]} + 2p(1-p) \times \frac{1}{2} \times \frac{W}{E[W]}) - (p \times 1)$, or $\text{cov}(z, W) = (1-p)/2$. Now recall that the term $E[wd]$ in the Price equation accounts for any differences in fidelity of transmission of the alleles, and since we assume each allele is transmitted without alteration, we have $E[wd] = 0$. Thus, overall, the Price equation becomes $\Delta p = (1-p)/2$. This gives the change in allele frequency over one generation, but we can also solve this recursion to obtain an equation for the frequency of allele D as a function of generation t. If the initial frequency is small, we obtain $p_t = 1 - (\frac{1}{2})^t$.

303 M. Gijzen, C. Ishmael, and S. D. Shrestha, "Epigenetic Control of Effectors in Plant Pathogens"; D. Qutob, B. P. Chapman, and M. Gijzen, "Transgenerational Gene Silencing Causes Gain of Virulence in a Plant Pathogen."

304 The pattern of inheritance for epialleles that we show in figure 9.1 is slightly different from that actually found. The experimental crosses show that all active alleles are silenced already in the F1 generation. We use the patterns in figure 1 to make the analysis of spread simpler. Modeling the spread of epialleles where the silencing happens already in the F1 generation only makes the results more extreme.

305 As before, we define z to be the frequency of the D epiallele *within an individual*. If we assume that mating is random then the type frequencies will be in Hardy-Weinberg equilibrium at the start of each generation; that is, for DD:DN:EE we have the frequencies $p^2 : 2p(1 - p) : (1 - p)^2$. Taking the fitness values specified in Box 9.2, the population mean fitness is pW as in Box 9.1. The covariance term is therefore identical to that from Box 9.1: $\text{cov}(z, W) = (1-p)/2$. The term $E[wd]$ in the Price equation accounts for any differences in fidelity of transmission of the epialleles. Unlike in Box 9.1, this will now be nonzero because the active epiallele in heterozygous individuals has some probability κ of being converted to the silenced epiallele. As a result, it can be shown that $E[wd] = \kappa(1-p)/2$. Thus, overall, the Price equation becomes $\Delta p = (1-p)/2 + \kappa(1-p)/2$ or $\Delta p = (1-p)(1+\kappa)/2$. As for Box 9.1 we can solve this recursion to obtain an equation for the frequency of allele D as a function of generation t. If the initial frequency is small, we obtain $p_t = 1 - (\frac{1-\kappa}{2})^t$.

306 Agrawal, Laforsch, and Tollrian, "Transgenerational Induction of Defences in Animals and Plants"; R. Poulin and F. Thomas, "Epigenetic Effects of Infection on the Phenotype of Host Offspring: Parasites Reaching across Host Generations."

307 R. Mostowy, J. Engelstadter, and M. Salathe, "Non-genetic Inheritance and the Patterns of Antagonistic Coevolution," explore some haploid cases of the Red Queen dynamic when there is nongenetic inheritance.

308 Darwin, *The Descent of Man*.

309 For a broad overview of courtship behavior and mate choice, see M. Andersson, *Sexual Selection*. For a description of the amazing morphology and courtship behavior of a peacock spider, see M. B. Girard, M. M. Kasumovic, and D. O. Elias, "Multi-modal Courtship in the Peacock Spider, *Maratus volans* (O.P.-Cambridge, 1874)." There are actually several species of peacocks and many species of peacock spiders, each with its own unique courtship display.

310 Darwin viewed mating as a cooperative process (see Darwin, *The Descent of Man*), but biologists now recognize that male-female interactions are rife with conflict (see G. Arnqvist and L. Rowe, *Sexual Conflict*). For example, in some species of birds, the most attractive males devote their efforts mainly to seeking extra-pair copulations, and tend to provide relatively poor paternal care. In insects, the most "well-armed" males may be quite harmful to females.

311 The conventional perspective also recognizes that phenotypic variation in condition can select for female preferences if low-condition males harbor parasites that they can transmit to females, or if they have poor-quality sperm. If such factors are important then there is no need to invoke any kind of benefit to offspring in explaining the evolution of female preferences. Females will clearly benefit by rejecting low-condition males in such systems. A more recent idea is that apparent female "preferences" can also result from sexual conflict. If some males can coerce females into mating, then females might mate with the most coercive males even though mating with such males reduces female fitness.

312 R. Bonduriansky and T. Day, "Nongenetic Inheritance and the Evolution of Costly Female Preference."

313 But not so high as to eliminate the correlation between a male's attractiveness and the quality of his offspring.

314 Females of many vertebrate species acquire their mate preferences socially, often by imprinting on the phenotype of their father. Unlike a genetic allele, such a learned preference for high-condition males could not be transmitted from a mother to her offspring. Rather, in a species where daughters acquired a preference for a male signal of high condition only if their father possessed such a signal, the frequency of the preference in the population would simply track the frequency of the male signal. In such a system, where female preference was also inherited nongenetically, sexual coevolution would therefore proceed quite differently from the situation that we modeled.

315 Crean, Adler, and Bonduriansky, "Seminal Fluid and Mate Choice: New Predictions."

316 This classic prediction has recently been confirmed in a natural population of New Zealand snails that includes both sexual and asexual types. The asexual snails increased in number at roughly twice the rate of the sexual ones (A. K. Gibson, L. F. Delph, and C. M. Lively. "The Two-Fold Cost of Sex: Experimental Evidence from a Natural System").

317 L Hadany and T. Beker, "On the Evolutionary Advantage of Fitness-Associated Recombination"; L. Hadany and S. P. Otto, "The Evolution of Condition-Dependent Sex in the Face of High Costs."

318 See L. M. Cosmides and J. Tooby, "Cytoplasmic Inheritance and Intragenomic Conflict"; A. L. Radzvilavicius, "Evolutionary Dynamics of Cytoplasmic Segregation and Fusion: Mitochondrial Mixing Facilitated the Evolution of Sex at the Origin of Eukaryotes"; J. C. Havird, M. D. Hall, and D. K. Dowling, "The Evolution of Sex: A New Hypothesis Based on Mitochondrial Mutational Erosion."

319 G. A. Parker, R. R. Baker, and V. G. Smith, "The Origin and Evolution of Gamete Dimorphism and the Male-Female Phenomenon."

320 https://www.australiazoo.com.au/our-animals/harriet/; http://www.smh.com.au/news/national/harriet-finally-withdraws-after-176-years/2006/06/23/1150845381649.html.

321 For an overview of these ideas, see K. A. Hughes and R. M. Reynolds, "Evolutionary and Mechanistic Theories of Aging."

322 This thought experiment is based on T. B. Kirkwood and M. R. Rose, "Evolution of Senescence: Late Survival Sacrificed for Reproduction."

323 T. M. Stubbs et al., "Multi-tissue DNA Methylation Age Predictor in Mouse"; S. Maegawa et al., "Caloric Restriction Delays Age-Related Methylation Drift."

324 L. P. Breitling et al., "Frailty Is Associated with the Epigenetic Clock but Not with Telomere Length in a German Cohort"; S. Horvath, "DNA Methylation Age of Human Tissues and Cell Types."

325 S. Horvath et al., "Obesity Accelerates Epigenetic Aging of Human Liver"; R. L. Simons et al., "Economic Hardship and Biological Weathering: The Epigenetics of Aging in a U.S. Sample of Black Women"; A. S. Zannas et al., "Lifetime Stress Accelerates Epigenetic Aging in an Urban, African American Cohort: Relevance of Glucocorticoid Signaling."

326 T. M. Stubbs et al., "Multi-tissue DNA Methylation Age Predictor in Mouse"; S. Maegawa et al., "Caloric Restriction Delays Age-Related Methylation Drift"; T. Wang et al., "Epigenetic Aging Signatures in Mice Livers Are Slowed by Dwarfism, Calorie Restriction, and Rapamycin Treatment"; J. J. Cole et al., "Diverse Interventions That Extend Mouse Lifespan Suppress Shared Age-Associated Epigenetic Changes at Critical Gene Regulatory Regions."

327 Fraga et al., "Epigenetic Differences Arise during the Lifetime of Monozygotic Twins."

328 E. Gilson and F. Magdinier, "Chromosomal Position Effect and Aging"; S. E. Johnstone and S. B. Baylin, "Stress and the Epigenetic Landscape: A Link to the Pathobiology of Human Disease?"; R. R. Kanherkar, N. Bhatia-Dey, and A. B. Csoka, "Epigenetics across the Human Lifespan"; P. Oberdoerffer and D. A. Sinclair, "The Role of Nuclear Architecture in Genomic Instability and Ageing"; R. M. Sapolsky, "Social Status and Health in Humans and Other Animals"; M. J. Sheriff, C. J. Krebs, and R. Boonstra, "The Sensitive Hare: Sublethal Effects of Predator Stress on Reproduction in Snowshoe Hares"; D. A. Sinclair and P. Oberdoerffer, "The Ageing Epigenome: Damaged Beyond Repair?"; L-Q. Cheng et al., "Epigenetic Regulation in Cell Senescence."

329 Carnes, Riesch, and Schlupp, "The Delayed Impact of Parental Age on Offspring Mortality in Mice"; Kern et al., "Decline in Offspring Viability as a Manifestation of Aging in *Drosophila melanogaster*"; Torres, Drummond, and Velando, "Parental Age and Lifespan Influence Offspring Recruitment: A Long-Term Study in a Seabird."

330 Gribble et al., "Maternal Caloric Restriction Partially Rescues the Deleterious Effects of Advanced Maternal Age on Offspring."

331 See Hercus and Hoffmann, "Maternal and Grandmaternal Age Influence Offspring Fitness in *Drosophila*"; the neriid evidence is as yet unpublished.

332 Sheriff, Krebs, and Boonstra, "The Sensitive Hare: Sublethal Effects of Predator Stress on Reproduction in Snowshoe Hares."

333 C. Zimmer, "What Is a Species?"

334 Pál and Miklós, "Epigenetic Inheritance, Genetic Assimilation, and Speciation." More recent reviews that examine the potential role of nongenetic inheritance in speciation include C. Lafon-Placette and C. Köhler, "Epigenetic Mechanisms of Postzygotic Reproductive Isolation in Plants"; D. W. Pfennig and M. R. Servedio, "The Role of Transgenerational Epigenetic Inheritance in Diversification and Speciation"; G. Smith and M. G. Ritchie, "How Might Epigenetics Contribute to Ecological Speciation?"

335 Klironomos, Berg, and Collins, "How Epigenetic Mutations Can Affect Genetic Evolution: Model and Mechanism."

336 T. A. Smith et al., "Epigenetic Divergence as a Potential First Step in Darter Speciation."

337 M. K. Skinner et al., "Epigenetics and the Evolution of Darwin's Finches."

338 Maynard Smith and Szathmáry, *The Major Transitions in Evolution.*

339 C. E. Juliano, S. Z. Swartz, and G. M. Wessel, "A Conserved Germline Multipotency Program."

340 A. Scheinfeld, "You and Heredity."

341 K. L. Jones et al., "Pattern of Malformation in Offspring of Chronic Alcoholic Mothers." An association between maternal alcoholism and developmental abnormalities in children had also been suggested a few years earlier by researchers in France (P. Lemoine et al., "Les enfants de parents alcooliques: Anomalies observées, a propos de 127 cas").

342 A. Streissguth et al., "Primary and Secondary Disabilities in Fetal Alcohol Syndrome."

343 Pauly, "How Did the Effects of Alcohol on Reproduction Become Scientifically Uninteresting?"; R. H. Warner and H. L. Rosett, "The Effects of Drinking on Offspring: A Historical Survey of the American and British Literature."

344 Warner and Rosett, "The Effects of Drinking on Offspring: A Historical Survey of the American and British Literature."

345 T. Wilson, *Distilled Spiritous Liquors the Bane of the Nation: Being Some Considerations Humbly Offer'd to the Lagislature, Part II.*

346 Warner and Rosett, "The Effects of Drinking on Offspring: A Historical Survey of the American and British Literature."

347 Georgios Papanicolaou, an immigrant from Greece, is most famous for inventing the Papanicolaou test, better known as the pap smear.

348 C. R Stockard and G. N. Papanicolaou, "Further Studies on the Modification of the Germ-Cells in Mammals: The Effect of Alcohol on Treated Guinea-Pigs and Their Descendants."

349 The potential for maternal drinking to harm the developing fetus was enthusiastically touted by some eugenicists as a "powerful agent of natural selection" for the control of reproduction by undesirable individuals and races (see E. M. Armstrong, *Conceiving Risk, Bearing Responsibility: Fetal Alcohol Syndrome and the Diagnosis of Moral Disorder*; J. Golden, *Message in a Bottle: The Making of Fetal Alcohol Syndrome*; Pauly,

"How Did the Effects of Alcohol on Reproduction Become Scientifically Uninteresting?"; Warner and Rosett, "The Effects of Drinking on Offspring: A Historical Survey of the American and British Literature").

350 Armstrong, *Conceiving Risk, Bearing Responsibility: Fetal Alcohol Syndrome and the Diagnosis of Moral Disorder*; Pauly, "How Did the Effects of Alcohol on Reproduction Become Scientifically Uninteresting?"

351 Stockard and Papanicolaou, "Further Studies on the Modification of the Germ-Cells in Mammals: The Effect of Alcohol on Treated Guinea-Pigs and Their Descendants."

352 Pauly, "How Did the Effects of Alcohol on Reproduction Become Scientifically Uninteresting?"

353 Hanson, "Modifications in the Albino Rat Following Treatment with Alcohol Fumes and X-Rays, and the Problem of Their Inheritance."

354 Armstrong, *Conceiving Risk, Bearing Responsibility: Fetal Alcohol Syndrome and the Diagnosis of Moral Disorder*; Pauly, "How Did the Effects of Alcohol on Reproduction Become Scientifically Uninteresting?"; Warner and Rosett, "The Effects of Drinking on Offspring: A Historical Survey of the American and British Literature."

355 "Effect of Alcoholism at Time of Conception."

356 H. W. Haggard, and E. M. Jellinek. *Alcohol Explored*, Garden City: Doubleday, 1942, quoted in Warner and Rosett, "The Effects of Drinking on Offspring: A Historical Survey of the American and British Literature."

357 Pauly, "How Did the Effects of Alcohol on Reproduction Become Scientifically Uninteresting?"; Warner and Rosett, "The Effects of Drinking on Offspring: A Historical Survey of the American and British Literature."

358 Scheinfeld, "You and Heredity."

359 E. L. Abel, *Fetal Alcohol Syndrome and Fetal Alcohol Effects*.

360 H. W. Haggard and E. M. Jellinek, *Alcohol Explored*, Garden City: Doubleday, 1942, quoted in Warner and Rosett, "The Effects of Drinking on Offspring: A Historical Survey of the American and British Literature."

361 Montagu, 1964, *Life before Birth* (114), cited in Abel, *Fetal Alcohol Syndrome and Fetal Alcohol Effects*.

362 E. P. Riley, M. A. Infante, and K. R. Warren, "Fetal Alcohol Spectrum Disorders: An Overview"; P. D. Sampson et al., "Incidence of Fetal Alcohol Syndrome and Prevalence of Alcohol-Related Neurodevelopmental Disorder."

363 For example, see A. Finegersh and G. E. Homanics, "Paternal Alcohol Exposure Reduces Alcohol Drinking and Increases Behavioral Sensitivity to Alcohol Selectively in Male Offspring," and references therein.

364 E. A. Mead and D. K. Sarkar, "Fetal Alcohol Spectrum Disorders and Their Transmission through Genetic and Epigenetic Mechanisms."

365 E. Jeyaratnam and S. Petrova, "Timeline: Key Events in the History of Thalidomide."

366 H. Sjöström and R. Nilsson, *Thalidomide and the Power of the Drug Companies*.

367 P. Knightley et al., *Suffer the Children: The Story of Thalidomide*.

368 J. Warkany, "Why I Doubted That Thalidomide Was the Cause of the Epidemic of Limb Defects of 1959 to 1961."

369 J. H. Kim and A. R. Scialli, "Thalidomide: The Tragedy of Birth Defects and the Effective Treatment of Disease."

370 P. H. Huang and W. G. McBride, "Interaction of [Glutarimide-2-^{14}C]Thalidomide with Rat Embryonic DNA In Vivo"; W. G. McBride and P. A. Read, "Thalidomide May Be a Mutagen."

371 D. Smithells, "Does Thalidomide Cause Second Generation Birth Defects?"

372 J. Laurance, "Experts Doubt Claims That Thalidomide Can Be Inherited."

373 D. J. Barker, "Fetal Origins of Coronary Heart Disease"; D.J.P. Barker, "The Fetal and Infant Origins of Adult Disease"; D.J.P. Barker, C. Osmond, and C. M. Law, "The Intrauterine and Early Postnatal Origins of Cardiovascular Disease and Chronic Bronchitis."

374 L. C. Schulz, "The Dutch Hunger Winter and the Developmental Origins of Health and Disease."

375 For example, see J. G. Eriksson, "The Fetal Origins Hypothesis—10 Years On"; K. S. Joseph and M. S. Kramer, "Review of the Evidence on Fetal and Early Childhood Antecendents of Adult Chronic Disease."

376 C. N. Hales and D.J.P. Barker, "Type 2 (Non-Insulin-Dependent) Diabetes Mellitus: The Thrifty Phenotype Hypothesis."

377 E. Archer, "The Childhood Obesity Epidemic as a Result of Nongenetic Evolution: The Maternal Resources."

378 Schulz, "The Dutch Hunger Winter and the Developmental Origins of Health and Disease."

379 M. T. Miller and K. Stromland, "Thalidomide: A Review, with a Focus on Ocular Findings and New Potential Uses."

380 T. J. Roseboom et al., "Effects of Prenatal Exposure to the Dutch Famine on Adult Disease in Later Life: An Overview."

381 W. H. Rooij, Sr. et al., "Prenatal Undernutrition and Cognitive Function in Late Adulthood"; Roseboom et al., "Effects of Prenatal Exposure to the Dutch Famine on Adult Disease in Later Life: An Overview."

382 S. A. Stanner and J. S. Yudkin, "Fetal Programming and the Leningrad Siege Study."

383 R. Painter et al., "Transgenerational Effects of Prenatal Exposure to the Dutch Famine on Neonatal Adiposity and Health in Later Life."

384 M. Veenendaal et al., "Transgenerational Effects of Prenatal Exposure to the 1944–45 Dutch Famine."

385 Dominguez-Salas et al., "Maternal Nutrition at Conception Modulates DNA Methylation of Human Metastable Epialleles"; Waterland et al., "Season of Conception in Rural Gambia Affects DNA Methylation at Putative Human Metastable Epialleles."

386 P. D. Gluckman et al., "Epigenetic Mechanisms That Underpin Metabolic and Cardiovascular Diseases"; H. Y. Zoghbi and A. L. Beaudet, "Epigenetics and Human Disease."

387 Pembrey et al., "Human Transgenerational Responses to Early-Life Experience: Potential Impact on Development, Health, and Biomedical Research."

388 Anway et al., "Epigenetic Transgenerational Actions of Endocrine Disruptors and Male Fertility"; Chen, Yan, and Duan, "Epigenetic Inheritance of Acquired Traits through Sperm RNAs and Sperm RNA Modifidations"; M. E. Pembrey et al., "Sex-Specific, Male-Line Transgenerational Responses in Humans."

389 J. C. Perry, L. K. Sirot, and S. Wigby, "The Seminal Symphony: How to Compose an Ejaculate."

390 Interesting evidence that factors in the seminal fluid can influence embryonic development comes from experiments on rodents in which the accessory glands of the male reproductive system (where various proteins and other components of the seminal fluid are synthesized) were removed. Although such males could still fertilize eggs, their offspring exhibited a range of developmental abnormalities (see J. J. Bromfield, "Seminal Fluid and Reproduction: Much More Than Previously Thought"; Crean, Adler, and Bonduriansky, "Seminal Fluid and Mate Choice: New Predictions").

391 M. Stoeckius, D. Grun, and N. Rajewsky, "Paternal RNA Contributions in the *Caenorhabditis elegans* Zygote."

392 Vojtech et al., "Exosomes in Human Semen Carry a Distinctive Repertoire of Small Non-coding RNAs with Potential Regulatory Functions."

393 Chen, Yan, and Duan, "Epigenetic Inheritance of Acquired Traits through Sperm RNAs and Sperm RNA Modifidations"; Eaton et al., "Roll over Weismann: Extracellular Vesicles in the Transgenerational Transmission of Environmental Effects."

394 Medical research shows that seminal fluid can increase the effectiveness of artificial reproductive technologies like IVF, and that increased maternal exposure to seminal fluid can reduce the incidence of pregnancy complications such as pre-eclampsia (see Crean, Adler, and Bonduriansky, "Seminal Fluid and Mate Choice: New Predictions").

395 Pembrey et al., "Sex-Specific, Male-Line Transgenerational Responses in Humans."

396 I. Donkin et al., "Obesity and Bariatric Surgery Drive Epigenetic Variation of Spermatozoa in Humans."

397 T. H. Chen, Y. H. Chiu, and B. J. Boucher, "Transgenerational Effects of Betel-Quid Chewing on the Development of the Metabolic Syndrome in the Keelung Community-Based Integrated Screening Program."

398 Reviewed in Pembrey et al., "Human Transgenerational Responses to Early-Life Experience: Potential Impact on Development, Health, and Biomedical Research."

399 Pembrey et al., "Human Transgenerational Responses to Early-Life Experience."

400 R. Yehuda et al., "Parental PTSD as a Vulnerability Factor for Low Cortisol Trait in Offspring of Holocaust Survivors."

401 R. Yehuda et al., "Holocaust Exposure Induced Intergenerational Effects on *FKBP5* Methylation."

402 Yehuda et al., "Parental PTSD as a Vulnerability Factor for Low Cortisol Trait in Offspring of Holocaust Survivors."

403 For example, some descendants of people who lived through the Chinese Cultural Revolution could be experiencing similar effects as the children of Holocaust survivors (see H. Gao, "A Scar on the Chinese Soul").

404 D. Crews et al., "Epigenetic Transgenerational Inheritance of Altered Stress Responses."

405 J. Zalasiewicz et al., "Are We Now Living in the Anthropocene?"

406 M. Latif and N. S. Keenlyside, "El Niño/Southern Oscillation Response to Global Warming."

407 J.E.M. Watson et al., "Catastrophic Declines in Wilderness Areas Undermine Global Environment Targets."

408 Bonduriansky, Crean, and Day, "The Implications of Nongenetic Inheritance for Evolution in Changing Environments." Also see R. E. O'Dea et al., "The Role of

Non-genetic Inheritance in Evolutionary Rescue: Epigenetic Buffering, Heritable Bet Hedging, and Epigenetic Traps."

409 S. Dey, S. R. Proulx, and H. Teotónio, "Adaptation to Temporally Fluctuating Environments by the Evolution of Maternal Effects."

410 See G. Gibson, *It Takes a Genome: How a Clash between Our Genes and Modern Life Is Making Us Sick.*

411 For example, see Gibson, *It Takes a Genome*; D. E. Lieberman, *The Story of the Human Body: Evolution, Health, and Disease.*

412 G. Cochran and H. Harpending, *The 10,000 Year Explosion: How Civilization Accelerated Human Evolution*; Sterelny, *The Evolved Apprentice.*

413 Although we tend to think of modern medicine as being very advanced, even ancient humans appear to have had some knowledge of the curative effects of antibiotics. For example, Nelson and colleagues (M. L. Nelson et al., "Mass Spectroscopic Characterization of Tetracycline in the Skeletal Remains of an Ancient Population from Sudanese Nubia 350–550 CE") have shown that 1,500-year-old human bones from a site in Africa show clear signs of high-dose tetracycline consumption. They speculate that members of this ancient population purposefully contaminated beer or other fermented beverages with material containing tetracycline-producing microbes, and then used the drink to treat various ailments. Even Neanderthals may have medicated themselves. Analysis of dental calculus from an individual with a dental abscess revealed traces of poplar, which contains salicylic acid, the pain-killing ingredient in aspirin (L. S. Weyrich et al., "Neanderthal Behaviour, Diet, and Disease Inferred from Ancient DNA in Dental Calculus").

414 https://www.nobelprize.org/nobel_prizes/medicine/laureates/1945/fleming-lecture.pdf.

415 J. W. Bigger, "Treatment of Staphylococcal Infections with Penicillin by Intermittent Sterilisation."

416 Many antibiotics, including penicillin, exert their effect by inhibiting the synthesis of bacterial cell walls during cell division. Therefore, if a bacterium goes dormant (that is, it stops reproducing), then it will not be affected by such drugs. Once the drug is removed, these dormant cells can reactivate and begin dividing (see N. Q. Balaban et al., "Bacterial Persistence as a Phenotypic Switch"; D. Shah et al., "Persisters: A Distinct Physiological State of *E. coli*").

417 E. Rotem et al., "Regulation of Phenotypic Variability by a Threshold-Based Mechanism Underlies Bacterial Persistence."

418 T. Bergmiller et al., "Biased Partitioning of the Multidrug Efflux Pump AcrAB-TolC Underlies Long-Lived Phenotypic Heterogeneity."

419 H. Easwaran, H. C. Tsai, and S. B. Baylin, "Cancer Epigenetics: Tumor Heterogeneity, Plasticity of Stem-Like States, and Drug Resistance."

420 Y. Ishino et al., "Nucleotide Sequence of the *iap* Gene, Responsible for Alkaline Phosphatase Isozyme Conversion in *Escherichia coli*, and Identification of the Gene Product."

421 After Nakata's discovery, similar findings were obtained for other bacterial species, and it was also noticed that these DNA repeats were separated by unique "spacer" segments of DNA. Moreover, in virtually all cases, the short DNA repeats were regularly interspersed with unique spacer sequences of a constant length, and these repetitive

patterns occurred in clusters. At this stage, very little was yet known about the biological function of CRISPR, but by 2005 the field of computational biology had advanced enough that researchers could carefully compare the genetic sequence of CRISPR from different species with virtually all known genetic sequences. Intriguingly, this revealed that the genetic sequences of the spacers in CRISPR were nearly identical to genetic sequences of viruses that infect bacteria. Bacteria, just like humans, are subject to viral infections and have evolved a variety of defense mechanisms. And so this discovery led to the speculation that CRISPR-Cas is some sort of bacterial immune system. Three papers published in 2005 began to lay the foundations for how CRISPR works. C. Pourcel, G. Salvignol, and G. Vergnaud, "CRISPR Elements in *Yersinia pestis* Acquire New Repeats by Preferential Uptake of Bacteriophage DNA, and Provide Additional Tools for Evolutionary Studies," documented that certain bacteria add spacers to the CRISPR system over time, while F. J. Mojica et al., "Intervening Sequences of Regularly Spaced Prokaryotic Repeats Derive from Foreign Genetic Elements," and A. Bolotin et al., "Clustered Regularly Interspaced Short Palindrome Repeats (CRISPRs) Have Spacers of Extrachromosomal Origin," documented that the spacers are very similar to genetic material of viral origin, and that their presence can influence the ability of viruses to infect the bacteria.

422 R. Barrangou et al., "CRISPR Provides Acquired Resistance against Viruses in Prokaryotes."

423 Interestingly, as we noted in chapter 3, the CRISPR system provides a clear example of the transmission of an environmentally induced trait, mediated by genetic material. A bacteria's infection experience alters the genetic composition of the spacers in its CRISPR system, and this environmentally induced change in the genome is then passed on to subsequent generations.

424 G. Gasiunas et al., "Cas9-crRNA Ribonucleoprotein Complex Mediates Specific DNA Cleavage for Adaptive Immunity in Bacteria"; M. Jinek et al., "Programmable Dual-RNA-Guided DNA Endonuclease in Adaptive Bacterial Immunity." The potential significance of the CRISPR system for genetic engineering is enormous, and at the time of writing, two groups are actively engaged in a patent dispute over who holds precedence in the development of this biomolecular technique. One group is based at the University of California, Berkeley, and the other is based at MIT and Harvard.

425 One of the reasons that this technology holds such enormous potential is that it is extremely easy to use compared with previous gene-editing techniques. Another reason is that the technology is relatively inexpensive and readily available to virtually anyone. See H. Ledford, "CRISPR, the Disruptor."

426 S. Reardon, "Welcome to the CRISPR Zoo."

427 D. Maddalo et al., "In Vivo Engineering of Oncogenic Chromosomal Rearrangements with the CRISPR/Cas9 System."

428 The phrase "tragedy of the commons" refers to a situation in which a resource that is available for exploitation by a group of individuals, each acting in its own self-interest, is depleted to the point where all individuals ultimately suffer. The idea was brought to widespread attention in the biological literature by Garrett Hardin in his paper "The Tragedy of the Commons."

429 Woody Guthrie wrote hundreds of songs, but he is perhaps best known for the song "This Land Is Your Land."

430 NIH 2015 statement on funding of research using gene-editing technologies in human embryos (https://www.nih.gov/about-nih/who-we-are/nih-director/statements/statement-nih-funding-research-using-gene-editing-technologies-human-embryos).

431 N. Wade, "Scientists Seek Moratorium on Edits to Human Genome That Could Be Inherited."

432 J. Gallagher, "Scientists Get 'Gene Editing' Go-Ahead."

433 Likewise, some scientists worry that, even without deliberate genetic modification, medical technologies will inevitably alter the human gene pool because relaxed natural selection on modern human populations will allow deleterious mutations to accumulate, with potentially catastrophic long-term consequences (see Lynch, "Mutation and Human Exceptionalism: Our Future Genetic Load").

434 See M. K. Skinner, "Environmental Stress and Epigenetic Transgenerational Inheritance"; "Endocrine Disruptor Induction of Epigenetic Transgenerational Inheritance of Disease."

435 J. Corrales et al., "Global Assessment of Bisphenol-A in the Environment."

436 J. C. Anderson, B. J. Park, and V. P. Palace, "Microplastics in Aquatic Environments: Implications for Canadian Ecosystems"; C. G. Avio, S. Gorbi, and F. Regoli, "Plastics and Microplastics in the Oceans: From Emerging Pollutants to Emerged Threat."

BIBLIOGRAPHY

Abel, E. L. *Fetal Alcohol Syndrome and Fetal Alcohol Effects*. New York: Plenum Press, 1993.

Abuddayeh, O. O., J. S. Gootenberg, S. Konermann, J. Joung, I. M. Slaymaker, D.B.T. Cox, S. Shmakov, et al. "C2c2 Is a Single-Component Programmable RNA-Guided RNA-Targeting CRISPR Effector." *Science* 353 (2016): aaf5573.

Agrawal, A. A., C. Laforsch, and R. Tollrian. "Transgenerational Induction of Defences in Animals and Plants." *Nature* 401 (1999): 60–63.

Aldrich, J. C., and K. A. Maggert. "Transgenerational Inheritance of Diet-Induced Genome Rearrangements in *Drosophila*." *PLoS Genetics* 11 (2015): e1005148.

Alonso-Magdalena, P., F. J. Rivera, and C. Guerrero-Bosagna. "Bisphenol-A and Metabolic Diseases: Epigenetic, Developmental, and Transgenerational Basis." *Environmental Epigenetics* 2 (2016): doi: 10.1093/eep/dvw022.

Amarger, V., A. Lecouillard, L. Ancellet, I. Grit, B. Castellano, P. Hulin, and P. Parnet. "Protein Content and Methyl Donors in Maternal Diet Interact to Influence the Proliferation Rate and Cell Fate of Neural Stem Cells in Rat Hippocampus." *Nutrients* 6 (2014): 4200–4217.

Anderson, J. C, B. J. Park, and V. P. Palace. "Microplastics in Aquatic Environments: Implications for Canadian Ecosystems." *Environmental Pollution* 218 (2016): 269–80.

Andersson, M. *Sexual Selection*. Princeton, NJ: Princeton University Press, 1994.

Anway, M. D., A. S. Cupp, M. Uzumucu, and M. K. Skinner. "Epigenetic Transgenerational Actions of Endocrine Disruptors and Male Fertility." *Science* 308 (2005): 1466–69.

Aplin, L. M., D. R. Farine, J. Morand-Ferron, A. Cockburn, A. Thornton, and B. C. Sheldon. "Experimentally Induced Innovations Lead to Persistent Culture via Conformity in Wild Birds." *Nature* 518 (2015): 538–41.

Archer, E. "The Childhood Obesity Epidemic as a Result of Nongenetic Evolution: The Maternal Resources." *Mayo Clinic Proceedings* 90 (2015): 77–92.

Armstrong, E. M. *Conceiving Risk, Bearing Responsibility: Fetal Alcohol Syndrome and the Diagnosis of Moral Disorder*. Baltimore: Johns Hopkins University Press, 2003.

Arnqvist, G., D. K. Dowling, P. Eady, L. Gay, T. Tregenza, M. Tuda, and D. J. Hosken. "Genetic Architecture of Metabolic Rate: Environment Specific Epistasis between Mitochondrial and Nuclear Genes in an Insect." *Evolution* 64 (2010): 3354–63.

Arnqvist, G., and L. Rowe. *Sexual Conflict*. Princeton, NJ: Princeton University Press, 2005.

Avio, C. G., S. Gorbi, and F. Regoli. "Plastics and Microplastics in the Oceans: From Emerging Pollutants to Emerged Threat." *Marine Environmental Research* 128 (2017): 2–11.

Avital, E., and E. Jablonka. *Animal Traditions: Behavioural Inheritance in Evolution*. Cambridge: Cambridge University Press, 2000.

Avramidou, E. V., I. V. Ganopoulos, A. G. Doulis, A. S. Tsaftaris, and F. A. Aravanopoulos. "Beyond Population Genetics: Natural Epigenetic Variation in Wild Cherry (*Prunus avium*)." *Tree Genetics and Genomes* 11 (2015): 95.

Badyaev, A. V. "Origin of the Fittest: Link between Emergent Variation and Evolutionary Change as a Critical Question in Evolutionary Biology." *Proceedings of the Royal Society B: Biological Sciences* 278 (2011): 1921–29.

Badyaev, A. V., and T. Uller. "Parental Effects in Ecology and Evolution: Mechanisms, Processes, and Implications." *Philosophical Transactions of the Royal Society B: Biological Sciences* 364 (2009): 1169–77.

Balaban, N. Q., J. Merrin, R. Chait, L. Kowalik, and S. Leibler. "Bacterial Persistence as a Phenotypic Switch." *Science* 305 (2004): 1622–25.

Baldanzi, S., R. Watson, C. D. McQuaid, G. Gouws, and F. Porri. "Epigenetic Variation among Natural Populations of the South African Sandhopper *Talorchestia capensis*." *Evolutionary Ecology* 31 (2016): 77–91.

Barker, D. J. "Fetal Origins of Coronary Heart Disease." *British Medical Journal* 311 (1995): 171–74.

Barker, D.J.P. "The Fetal and Infant Origins of Adult Disease." *British Medical Journal* 301 (1990): 1111.

Barker, D.J.P., C. Osmond, and C. M. Law. "The Intrauterine and Early Postnatal Origins of Cardiovascular Disease and Chronic Bronchitis." *Journal of Epidemiology and Community Health* 43 (1989): 237–40.

Barrangou, R., C. Fremaux, H. Deveau, M. Richards, P. Boyaval, S. Moineau, D. A. Romero, and P. Horvath. "CRISPR Provides Acquired Resistance against Viruses in Prokaryotes." *Science* 315 (2007): 1709–12.

Bateson, W. *William Bateson, F.R.S., Naturalist: His Essays and Addresses Together with a Short Account of His Life by Beatrice Bateson*. Cambridge: Cambridge University Press, 1928.

Beisson, J. "Preformed Cell Structure and Cell Heredity." *Prion* 2 (2008): 1–8.

Beisson, J., and T. M. Sonneborn. "Cytoplasmic Inheritance of the Organization of the Cell Cortex in *Paramecium aurelia*." *Proceedings of the National Academy of Sciences USA* 53 (1965): 275–82.

Benkemoun, L., and S. J. Saupe. "Prion Proteins as Genetic Material in Fungi." *Fungal Genetics and Biology* 43 (2006): 789–803.

Bergmiller, T., A.M.C. Andersson, K. Tomasek, E. Balleza, D. J. Kiviet, R. Hauschild, G. Tkačik, and C. C. Guet. "Biased Partitioning of the Multidrug Efflux Pump AcrAB-TolC Underlies Long-Lived Phenotypic Heterogeneity." *Science* 356 (2017): 311–15.

Bhandari, R. K., F. S. vom Saal, and D. E. Tillitt. "Transgenerational Effects from Early Developmental Exposures to Bisphenol A or 17a-Ethinylestradiol in Medaka, *Oryzias latipes*." *Scientific Reports* 5 (2015): 9303.

Bigger, J. W. "Treatment of Staphylococcal Infections with Penicillin by Intermittent Sterilisation." *The Lancet* 244 (1944): 497–500.

Blewitt, M. E., N. K. Vickaryous, A. Paldi, H. Koseki, and E. Whitelaw. "Dynamic Reprogramming of DNA Methylation at an Epigenetically Sensitive Allele in Mice." *PLoS Genetics* 2 (2006): e49.

Bolotin, A., B. Quinquis, A. Sorokin, and S. D. Ehrlich. "Clustered Regularly Interspaced Short Palindrome Repeats (CRISPRs) Have Spacers of Extrachromosomal Origin." *Microbiology* 151 (2005): 2551–61.

Bonduriansky, R. "The Ecology of Sexual Conflict: Background Mortality Can Modulate the Effects of Male Manipulation on Female Fitness." *Evolution* 68 (2014): 595–604.

———. "Rethinking Heredity, Again." *Trends in Ecology and Evolution* 27 (2012): 330–36.

Bonduriansky, R., and A. J. Crean. "What Are Parental Condition-Transfer Effects and How Can They Be Detected?" *Methods in Ecology and Evolution* (2017): DOI 10.1111/2041-210X.12848.

Bonduriansky, R., A. J. Crean, and T. Day. "The Implications of Nongenetic Inheritance for Evolution in Changing Environments." *Evolutionary Applications* 5 (2012): 192–201.

Bonduriansky, R., and T. Day. "Nongenetic Inheritance and Its Evolutionary Implications." *Annual Review of Ecology, Evolution, and Systematics* 40 (2009): 103–25.

———. "Nongenetic Inheritance and the Evolution of Costly Female Preference." *Journal of Evolutionary Biology* 26 (2013): 76–87.

Bonduriansky, R., and M. Head. "Maternal and Paternal Condition Effects on Offspring Phenotype in *Telostylinus angusticollis* (Diptera : Neriidae)." *Journal of Evolutionary Biology* 20 (2007): 2379–88.

Bonduriansky, R., A. Runagall-McNaull, and A. J. Crean. "The Nutritional Geometry of Parental Effects: Maternal and Paternal Macronutrient Consumption and Offspring Phenotype in a Neriid Fly." *Functional Ecology* 30 (2016): 1675–86.

Bornens, M. "Organelle Positioning and Cell Polarity." *Nature Reviews Molecular and Cell Biology* 9 (2008): 874–86.

Boto, L. "Horizontal Gene Transfer in Evolution: Facts and Challenges." *Proceedings of the Royal Society B: Biological Sciences* 277 (2010): 819–27.

Boveri, Th. "An Organism Produced Sexually without Characteristics of the Mother." *American Naturalist* 27 (1893): 222–32.

Bowler, P. J. *The Mendelian Revolution: The Emergence of Hereditarian Concepts in Modern Science and Society*. London: Athlone Press, 1989.

Boyd, R., and P. J. Richerson. *Culture and the Evolutionary Process*. Chicago: University of Chicago Press, 1985.

Breitling, L. P., K-U. Saum, L. Perna, B. Schöttker, B. Holleczek, and H. Brenner. "Frailty Is Associated with the Epigenetic Clock but Not with Telomere Length in a German Cohort." *Clinical Epigenetics* 8 (2016): 21.

Bremer, J., F. Baumann, C. Tiberi, C. Weissig, H. Fischer, P. Schwarz, A. D. Steele, et al. "Axonal Prion Protein Is Required for Peripheral Myelin Maintenance." *Nature Neuroscience* 13 (2010): 310–18.

Bromfield, J. J. "Seminal Fluid and Reproduction: Much More Than Previously Thought." *Journal of Assisted Reproduction and Genetics* 31 (2014): 627–36.

Calhoun, K. C., E. Padilla-Banks, W. N. Jefferson, L. Liu, K. E. Gerrish, S. L. Young, C. E. Wood, et al. "Bisphenol A Exposure Alters Developmental Gene Expression in the Fetal Rhesus Macaque Uterus." *PLoS One* 9 (2014): e85894.

Camus, M. F., D. J. Clancy, and D. K. Dowling. "Mitochondria, Maternal Inheritance, and Male Aging." *Current Biology* 22 (2012): 1717–21.
Camus, M. F., J.B.W. Wolf, E. H. Morrow, and D. K. Dowling. "Single Nucleotides in the mtDNA Sequence Modify Mitochondrial Molecular Function and Are Associated with Sex-Specific Effects on Fertility and Aging." *Current Biology* 25 (2015): 2717–22.
Caprette, C. L., M.S.Y. Lee, R. Shine, A. Mokany, and J. F. Downhower. "The Origin of Snakes (Serpentes) as Seen through Eye Anatomy." *Biological Journal of the Linnean Society* 81 (2004): 469–82.
Carey, N. *The Epigenetics Revolution: How Modern Biology Is Rewriting Our Understanding of Genetics, Disease, and Inheritance*. New York: Columbia University Press, 2012.
Carnes, B. A., R. Riesch, and I. Schlupp. "The Delayed Impact of Parental Age on Offspring Mortality in Mice." *Journals of Gerontology: Series A, Biological Sciences and Medical Sciences* 67A (2012): 351–57.
Cavalier-Smith, T. "The Membranome and Membrane Heredity in Development and Evolution." In *Organelles, Genomes, and Eukaryote Phylogeny: An Evolutionary Synthesis in the Age of Genomics*, edited by R. P. Hirt and D. S. Horner, 335–52. Boca Raton, FL: CRC Press, 2004.
Cavalli-Sforza, L. L., and M. W. Feldman. *Cultural Transmission and Evolution: A Quantitative Approach*. Princeton, NJ: Princeton University Press, 1981.
Chandler, V., and M. Alleman. "Paramutation: Epigenetic Instructions Passed across Generations." *Genetics* 178 (2008): 1839–44.
Chen, Q., Y. Menghong, C. Zhonghong, X. Li, Y. Zhang, J. Shi, G-h. Feng, et al. "Sperm tsRNAs Contribute to Intergenerational Inheritance of an Acquired Metabolic Disorder." *Science* 351 (2016): 397–400.
Chen, Q., W. Yan, and E. Duan. "Epigenetic Inheritance of Acquired Traits through Sperm RNAs and Sperm RNA Modifications." *Nature Reviews Genetics* 17 (2016): 733–43.
Chen, T. H., Y. H. Chiu, and B. J. Boucher. "Transgenerational Effects of Betel-Quid Chewing on the Development of the Metabolic Syndrome in the Keelung Community-Based Integrated Screening Program." *American Journal of Clinical Nutrition* 83 (2006): 688–92.
Cheng, L-Q., Z-Q. Zhang, H-Z. Chen, and D-P. Liu. "Epigenetic Regulation in Cell Senescence." *Journal of Molecular Medicine* (2017): DOI 10.1007/s00109-017-1581-x.
Chetverin, A. B. "Can a Cell Be Assembled from Its Constituents?" *Paleontological Journal* 44 (2010): 715–27.
Chi, K. R. "The RNA Code Comes into Focus." *Nature* 542 (2017): 503–6.
Cochran, G., and H. Harpending. *The 10,000 Year Explosion: How Civilization Accelerated Human Evolution*. New York: Basic Books, 2010.
Cole, J. J., N. A. Robertson, M. I. Rather, J. P. Thomson, T. McBryan, D. Sproul, T. Wang, C. Brock, W. Clark, T. Ideker, R. R. Meehan, R. A. Miller, H. M. Brown-Borg, and P. D. Adams. "Diverse Interventions That Extend Mouse Lifespan Suppress Shared Age-Associated Epigenetic Changes at Critical Gene Regulatory Regions." *Genome Biology* 18 (2017): 58.
Cook, G. M. "Neo-Lamarckian Experimentalism in America: Origins and Consequences." *Quarterly Review of Biology* 74 (1999): 417–37.
Corrales, J., L. A. Kristofco, W. B. Steele, B. S. Yates, C. S. Breed, W. Spencer, and B. W. Brooks. "Global Assessment of Bisphenol A in the Environment." *Dose-Response* 13 (2015): 1559325815598308.

Cosmides, L. M., and J. Tooby. "Cytoplasmic Inheritance and Intragenomic Conflict." *Journal of Theoretical Biology* 89 (1981): 83–129.
Cowley, J. J., and R. D. Griesel. "The Effect on Growth and Behaviour of Rehabilitating First and Second Generation Low Protein Rats." *Animal Behaviour* 14 (1966): 506–17.
Crean, A. J., M. I. Adler, and R. Bonduriansky. "Seminal Fluid and Mate Choice: New Predictions." *Trends in Ecology and Evolution* 31 (2016): 253–55.
Crean, A. J., and R. Bonduriansky. "What Is a Paternal Effect?" *Trends in Ecology and Evolution* 29 (2014): 554–59.
Crean, A. J., A. M. Kopps, and R. Bonduriansky. "Revisiting Telegony: Offspring Inherit an Acquired Characteristic of Their Mother's Previous Mate." *Ecology Letters* 17 (2014): 1545–52.
Crews, D., R. Gillette, S. V. Scarpino, M. Manikkam, M. I. Savenkova, and M. K. Skinner. "Epigenetic Transgenerational Inheritance of Altered Stress Responses." *Proceedings of the National Academy of Sciences USA* 109 (2012): 9143–48.
Crick, F. H. C. "The Croonian Lecture: The Genetic Code." *Proceedings of the Royal Society of London B: Biological Sciences* 167 (1966): 331–47.
Cropley, J. E., T.H.Y. Dant, D.I.K. Martin, and C. M. Suter. "The Penetrance of an Epigenetic Trait in Mice Is Progressively Yet Reversibly Increased by Selection and Environment." *Proceedings of the Royal Society of London B: Biological Sciences* 279 (2012): 2347–53.
Cropley, J. E., C. M. Suter, E. B. Beckman, and D. I. Martin. "Germ-Line Epigenetic Modification of the Murine A^{vy} Allele by Nutritional Supplementation." *Proceedings of the National Academy of Sciences USA* 103 (2006): 17308–12.
Cubas, P., C. Vincent, and E. Coen. "An Epigenetic Mutation Responsible for Natural Variation in Floral Symmetry." *Nature* 401 (1999): 157–61.
Curley, J. P., R. Mashoodh, and F. A. Champagne. "Epigenetics and the Origins of Paternal Effects." *Hormones and Behavior* 59 (2011): 306–14.
Danchin, E. "Avatars of Information: Towards an Inclusive Evolutionary Synthesis." *Trends in Ecology and Evolution* 28 (2013): 351–58.
Danchin, E., A. Charmantier, F. A. Champagne, A. Mesoudi, B. Pujol, and S. Blanchet. "Beyond DNA: Integrating Inclusive Inheritance into an Extended Theory of Evolution." *Nature Reviews Genetics* 12 (2011): 475–86.
Danchin, E., L-A. Giraldeau, T. J. Valone, and R. H. Wagner. "Public Information: From Noisy Neighbors to Cultural Evolution." *Science* 305 (2004): 487–91.
Danchin, E., and R. H. Wagner. "Inclusive Heritability: Combining Genetic and Nongenetic Information to Study Animal Behavior and Culture." *Oikos* 119 (2010): 210–18.
Darwin, C. R. *The Descent of Man*. London: John Murray, 1871.
———. *On the Origin of Species*. 2nd ed. London: John Murray, 1859.
———. *The Variation of Animals and Plants under Domestication*. 2nd ed. Vol. 1. London: John Murray, 1875.
Daxinger, L., and E. Whitelaw. "Understanding Transgenerational Epigenetic Inheritance via the Gametes in Mammals." *Nature Reviews Genetics* 13 (2012): 152–62.
Day, T., and R. Bonduriansky. "A Unified Approach to the Evolutionary Consequences of Genetic and Nongenetic Inheritance." *American Naturalist* 178 (2011): E18–E136.
DeJong-Lambert, W. *The Cold War Politics of Genetic Research: An Introduction to the Lysenko Affair*. New Studies in the History and Philosophy of Science and Technology. Dordrecht: Springer, 2012.

Devanapally, S., S. Ravikumar, and A. M. Jose. "Double-Stranded RNA Made in *C. elegans* Neurons Can Enter the Germline and Cause Transgenerational Gene Silencing." *Proceedings of the National Academy of Sciences USA* 112 (2015): 2133–38.

Dey, S., S. R. Proulx, and H. Teotónio. "Adaptation to Temporally Fluctuating Environments by the Evolution of Maternal Effects." *PloS Biology* 14 (2016): e1002388.

Dias, B. G., and K. J. Ressler. "Parental Olfactory Experience Influences Behaviour and Neural Structure in Subsequent Generations." *Nature Neuroscience* 17 (2014): 89–96.

Dickins, B., and Q. Rahman. "The Extended Evolutionary Synthesis and the Role of Soft Inheritance in Evolution." *Proceedings of the Royal Society B: Biological Sciences* 279 (2012): 2913–21.

Dickins, T. E., and B.J.A. Dickins. "Mother Nature's Tolerant Ways: Why Non-genetic Inheritance Has Nothing to Do with Evolution." *New Ideas in Psychology* 26 (2008): 41–54.

Dobzhansky, T. *Genetics and the Origin of Species*. New York: Columbia University Press, 1951.

———. *Genetics of the Evolutionary Process*. New York: Columbia University Press, 1970.

Dominguez-Salas, P., S. E. Moore, M. S. Baker, A. W. Bergen, S. E. Cox, R. A. Dyer, A. J. Fulford, et al. "Maternal Nutrition at Conception Modulates DNA Methylation of Human Metastable Epialleles." *Nature Communications* 5 (2014): 3746.

Donkin, I., S. Versteyhe, L. R. Ingerslev, K. Qian, M. Mechta, L. Nordkap, B. Mortensen, et al. "Obesity and Bariatric Surgery Drive Epigenetic Variation of Spermatozoa in Humans." *Cell Metabolism* 23 (2016): 369–78.

Dubin, M. J., P. Zhang, D. Meng, M.-S. Remigereau, E. J. Osborne, F. P. Casale, P. Drewe, et al. "DNA Methylation in *Arabidopsis* Has a Genetic Basis and Shows Evidence of Local Adaptation." *eLife* 4 (2015): e05255.

Dunn, L. C., and T. Dobzhansky. *Heredity, Race, and Society*. New York: New American Library of World Literature, 1946.

Dussourd, D. E., K. Ubik, C. Harvis, J. Resch, J. Meinwald, and T. Eisner. "Biparental Defensive Endowment of Eggs with Acquired Plant Alkaloid in the Moth *Utetheisa ornatrix*." *Proceedings of the National Academy of Sciences USA* 85 (1988): 5992–96.

Easwaran, H., H. C. Tsai, and S. B. Baylin. "Cancer Epigenetics: Tumor Heterogeneity, Plasticity of Stem-Like States, and Drug Resistance." *Molecular Cell* 54 (2014): 716–27.

Eaton, S. A., N. Jayasooriah, M. E. Buckland, D. I. Martin, J. E. Cropley, and C. M. Suter. "Roll over Weismann: Extracellular Vesicles in the Transgenerational Transmission of Environmental Effects." *Epigenomics* 7 (2015): 1165–71.

"Effect of Alcoholism at Time of Conception." *Journal of the American Medical Association* 146 (1946): 419.

Eisenberg, D.T.A. "Inconsistent Inheritance of Telomere Length (TL): Is Offspring TL More Strongly Correlated with Maternal or Paternal TL?" *European Journal of Human Genetics* 22 (2014): 8–9.

Elsworth, J. D., J. D. Jentsch, S. M. Groman, R. H. Roth, E. D. Redmond Jr., and C. Leranth. "Low Circulating Levels of Bisphenol-A Induce Cognitive Deficits and Loss of Asymmetric Spine Synapses in Dorsolateral Prefrontal Cortex and Hippocampus of Adult Male Monkeys." *Journal of Comparative Neurology* 523 (2015): 1248–57.

Epel, E. S., E. H. Blackburn, J. Lin, F. S. Dhabhar, N. E. Adler, J. D. Morrow, and R. M. Cawthorn. "Accelerated Telomere Shortening in Response to Life Stress." *Proceedings of the National Academy of Sciences USA* 101 (2004): 17312–15.

Eriksson, J. G. "The Fetal Origins Hypothesis—10 Years On." *British Medical Journal* 330 (2005): 1096–97.

Erlich, Y., and D. Zielinski. "DNA Fountain Enables a Robust and Efficient Storage Architecture." *Science* 355 (2017): 950–54.

Ernst, U. R., M. B. Van Hiel, G. Depuydt, B. Boerjan, A. De Loof, and L. Shoofs. "Epigenetics and Locust Life Phase Transitions." *Journal of Experimental Biology* 218 (2015): 88–99.

Feinberg, A. P., and R. A. Irizarry. "Stochastic Epigenetic Variation as a Driving Force of Development, Evolutionary Adaptation, and Disease." *Proceedings of the National Academy of Sciences USA* 107 (2010): 1757–64.

Finegersh, A., and G. E. Homanics. "Paternal Alcohol Exposure Reduces Alcohol Drinking and Increases Behavioral Sensitivity to Alcohol Selectively in Male Offspring." *PLoS One* 9 (2014): e99078.

Fioriti, L., C. Myers, Y-Y. Huang, X. Li, J. S. Stephan, P. Trifilieff, L. Colnaghi, et al. "The Persistence of Hippocampal-Based Memory Requires Protein Synthesis Mediated by the Prion-Like Protein CPEB3." *Neuron* 86 (2015): 1433–48.

Firestein, S. *Ignorance*. Oxford: Oxford University Press, 2012.

Ford, E. B. *Mendelism and Evolution*. London: Methuen, 1931.

Fraga, M. F., E. Bellestar, M. F. Paz, S. Ropero, F. Setien, M. L. Bellestar, D. Heine-Suner, et al. "Epigenetic Differences Arise during the Lifetime of Monozygotic Twins." *Proceedings of the National Academy of Sciences USA* 102 (2005): 10604–9.

Francis, R. C. *Epigenetics: How Environment Shapes Our Genes*. New York: W. W. Norton, 2011.

Frank, S. A. "George Price's Contributions to Evolutionary Genetics." *Journal of Theoretical Biology* 175 (1995): 373–88.

Fraser, R., and C-J. Lin. "Epigenetic Reprogramming of the Zygote in Mice and Men: On Your Marks, Get Set, Go!" *Reproduction in Domestic Animals* 152 (2016): R211–R22.

Frazier, C. R. M., B. C. Trainor, C. J. Cravens, T. K. Whitney, and C. A. Marler. "Paternal Behavior Influences Development of Aggression and Vasopressin Expression in Male California Mouse Offspring." *Hormones and Behavior* 50 (2006): 699–707.

Furrow, R.E. "Epigenetic Inheritance, Epimutation, and the Response to Selection." *PLoS One* 9 (2014): e101559.

Furrow, R. E., F. B. Christiansen, and M. W. Feldman. "Environment-Sensitive Epigenetics and the Heritability of Complex Diseases." *Genetics* 189 (2011): 1377–87.

Futuyma, D. J. "Can Modern Evolutionary Theory Explain Macroevolution?" In *Macroevolution: Explanation, Interpretation, and Evidence*, edited by E. Serrelli and N. Gontier, 29–85. Cham, Switzerland: Springer International Publishing, 2015.

Gallagher, J. "Scientists Get 'Gene Editing' Go-Ahead." *BBC Health* (2016).

Galton, F. "Experiments in Pangenesis, by Breeding from Rabbits of a Pure Variety, into Whose Circulation Blood Taken from Other Varieties Had Previously Been Largely Transfused." *Proceedings of the Royal Society of London* 19 (1871): 393–410.

———. "Hereditary Character and Talent." *Macmillan's Magazine* 12 (1865): 157–66.

———. "Hereditary Improvement." *Fraser's Magazine* January (1873): 116–30.

Gao, H. "A Scar on the Chinese Soul." *New York Times*, January 18, 2017.

Gapp, K., A. Jawaid, P. Sarkies, J. Bohacek, P. Pelczar, J. Prados, L. Farinelli, E. Miska, and I. M. Mansuy. "Implication of Sperm RNAs in Transgenerational Inheritance of the Effects of Early Trauma in Mice." *Nature Neuroscience* 17 (2016): 667–69.

Garcia-Gonzalez, F., and D. K. Dowling, "Transgenerational Effects of Sexual Interactions and Sexual Conflict: Non-Sires Boost the Fecundity of Females in the Following Generation." *Biology Letters* 11 (2015): 20150067.

Gardner, A. "The Price Equation." *Current Biology* 18 (2008): R198–R202.

Garver, K. L., and B. Garver. "Eugenics: Past, Present, and the Future." *American Journal of Human Genetics* 49 (1991): 1109–18.

Gasiunas, G., R. Barrangou, P. Horvath, and V. Siksnys. "Cas9-crRNA Ribonucleoprotein Complex Mediates Specific DNA Cleavage for Adaptive Immunity in Bacteria." *Proceedings of the National Academy of Sciences USA* 109 (2012): E2579–E86.

Geoghegan, J. L., and H. G. Spencer. "Exploring Epiallele Stability in a Population-Epigenetic Model." *Theoretical Population Biology* 83 (2012): 136–44.

———. "Population-Epigenetic Models of Selection." *Theoretical Population Biology* 81 (2012): 232–42.

Geoghegan, J. L., and H. G. Spencer. "The Evolutionary Potential of Paramutation: A Population-Epigenetic Model." *Theoretical Population Biology* 88 (2013): 9–19.

Gibson, A. K., L. F. Delph, and C. M. Lively. "The Two-Fold Cost of Sex: Experimental Evidence from a Natural System." *Evolution Letters* 1 (2017): 6–15.

Gibson, D. G., J. I. Glass, C. Lartigue, V. N. Noskov, R-Y. Chuang, M. A. Algire, G. A. Benders, et al. "Creation of a Bacterial Cell Controlled by a Chemically Synthesized Genome." *Science* 329 (2010): 52–56.

Gibson, G. *It Takes a Genome: How a Clash between Our Genes and Modern Life Is Making Us Sick*. Upper Saddle River, NJ: FT Press Science, 2009.

Gijzen, M., C. Ishmael, and S. D. Shrestha. "Epigenetic Control of Effectors in Plant Pathogens." *Frontiers in Plant Science* 5 (2014): 638.

Gilbert, S. F., J. Sapp, and A. I. Tauber. "A Symbiotic View of Life: We Have Never Been Individuals." *Quarterly Review of Biology* 87 (2012): 325–41.

Gilson, E., and F. Magdinier. "Chromosomal Position Effect and Aging." In *Epigenetics of Aging*, edited by T. Tollefsbol, 151–76. New York: Springer, 2010.

Girard, M. B., M. M. Kasumovic, and D. O. Elias. "Multi-modal Courtship in the Peacock Spider, *Maratus volans* (O.P.-Cambridge, 1874)." *PLoS One* 6 (2011): e25390.

Gleick, J. *Genius: The Life and Science of Richard Feynman*. New York: Vintage Books, 1993.

Gluckman, P. D., M. A. Hanson, T. Buklijas, F. M. Low, and A. S. Beedle. "Epigenetic Mechanisms That Underpin Metabolic and Cardiovascular Diseases." *Nature Reviews Endocrinology* 5 (2009): 401–8.

Godfrey-Smith, P. *Darwinian Populations and Natural Selection*. Oxford: Oxford University Press, 2009.

———. "Is It a Revolution?" *Biology and Philosophy* 22 (2007): 429–37.

Gokhman, D., E. Lavi, K. Prufer, M. F. Fraga, J. A. Riancho, J. Kelso, S. Paabo, E. Meshorer, and L. Carmel. "Reconstructing the DNA Methylation Maps of the Neandertal and the Denisovan." *Science* 344 (2014): 523–27.

Gokhman, D., A. Malul, and L. Carmel. "Inferring Past Environments from Ancient Epigenomes." *Molecular Biology and Evolution* 34 (2017): 2429–38.

Golden, J. *Message in a Bottle: The Making of Fetal Alcohol Syndrome*. Cambridge, MA: Harvard University Press, 2006.

Gordon, L. B., F. G. Rothman, C. Lopez-Otin, and T. Misteli. "Progeria: A Paradigm for Translational Medicine." *Cell* 156 (2014): 400–407.

Grandjean, V., S. Fourré, D. Fernandes De Abreu, M-A. Derieppe, J-J. Remy, and M. Rassoulzadegan. "RNA-Mediated Paternal Heredity of Diet-Induced Obesity and Metabolic Disorders." *Scientific Reports* 5 (2015): 18193.

Green, R. E., J. Krause, A. W. Briggs, T. Maricic, U. Stenzel, M. Kircher, M. Patterson, et al. "A Draft Sequence of the Neandertal Genome." *Science* 328 (2010): 710–22.

Greer, E. L., T. J. Maures, A. G. Hauswirth, E. M. Green, D. S. Leeman, G. S. Maro, S. Han, et al. "Members of the H3K4 Trimethylation Complex Regulate Lifespan in a Germline-Dependent Manner in *C. elegans*." *Nature* 466 (2010): 383–87.

Greer, E. L., T. J. Maures, D. Ucar, A. G. Hauswirth, E. Mancini, J. P. Lim, B. A. Benayoun, Y. Shi, and A. Brunet. "Transgenerational Epigenetic Inheritance of Longevity in *Caenorhabditis elegans*." *Nature* 479 (2011): 365–71.

Gribble, K. E., G. Jarvis, M. Bock, and D.B.M. Welch. "Maternal Caloric Restriction Partially Rescues the Deleterious Effects of Advanced Maternal Age on Offspring." *Aging Cell* 13 (2014): 623–30.

Griffiths, P. E., and R. D. Gray. "Developmental Systems and Evolutionary Explanation." *Journal of Philosophy* 91 (1994): 277–304.

Grimes, G. W. "Pattern Determination in Hypotrich Ciliates." *American Zoologist* 22 (1982): 35–46.

Hackett, J. A., J. J. Zylicz, and A. Surani. "Parallel Mechanisms of Epigenetic Reprogramming in the Germline." *Trends in Genetics* 28 (2012): 164–74.

Hadany, L., and T. Beker. "On the Evolutionary Advantage of Fitness-Associated Recombination." *Genetics* 165 (2003): 2167–79.

Hadany, L., and S. P. Otto. "The Evolution of Condition-Dependent Sex in the Face of High Costs." *Genetics* 176 (2007): 1713–27.

Haig, D. "The Kinship Theory of Genomic Imprinting." *Annual Review of Ecology and Systematics* 31 (2000): 9–32.

Haig, D. "Weismann Rules! OK? Epigenetics and the Lamarckian Temptation." *Biology and Philosophy* 22 (2007): 415–28.

Haldane, J.B.S. *Adventures of a Biologist*. New York: Harper and Brothers, 1937.

Haldane, J.B.S., and J. Huxley. *Animal Biology*. Oxford: Clarendon Press, 1934.

Hales, C. N., and D.J.P. Barker. "Type 2 (Non-Insulin-Dependent) Diabetes Mellitus: The Thrifty Phenotype Hypothesis." *Diabetologia* 35 (1992): 595–601.

Halfmann, R., S. Alberti, and S. Lindquist. "Prions, Protein Homeostasis, and Phenotypic Diversity." *Trends in Cell Biology* 20 (2010): 125–33.

Halfmann, R., and S. Lindquist. "Epigenetics in the Extreme: Prions and the Inheritance of Environmentally Acquired Traits." *Science* 330 (2010): 629–32.

Hanson, F. B. "Modifications in the Albino Rat Following Treatment with Alcohol Fumes and X-Rays, and the Problem of Their Inheritance." *Proceedings of the American Philosophical Society* 62 (1923): 301–10.

Harari, Y. N. *Homo Deus: A Brief History of Tomorrow*. London: Harvill Secker, 2016.

Hardin, G. "The Tragedy of the Commons." *Science* 162 (1968): 1243–48.

Haring, M., R. Bader, M. Louwers, A. Schwabe, R. van Driel, and M. Stam. "The Role of DNA Methylation, Nucleosome Occupancy and Histone Modifications in Paramutation." *Plant Journal* 63 (2010): 366–78.

Harman, O. *The Price of Altruism*. London: Bodley Head, 2010.

Havird, J. C., M. D. Hall, and D. K. Dowling. "The Evolution of Sex: A New Hypothesis Based on Mitochondrial Mutational Erosion." *Bioessays* 37 (2015): 951–58.

Helanterä, H., and T. Uller. "The Price Equation and Extended Inheritance." *Philosophy and Theory in Biology* 2 (2010): e101.

Hendry, A. P. *Eco-Evolutionary Dynamics*. Princeton, NJ: Princeton University Press, 2017.

Hercus, M. J., and A. A. Hoffmann. "Maternal and Grandmaternal Age Influence Offspring Fitness in *Drosophila*." *Proceedings of the Royal Society of London B: Biological Sciences* 267 (2000): 2105–10.

Herrera, C. M., and P. Bazaga. "Untangling Individual Variation in Natural Populations: Ecological, Genetic, and Epigenetic Correlates of Long-Term Inequality in Herbivory." *Molecular Ecology* 20 (2011): 1675–88.

Hirsch, S., R. Baumberger, and U. Grossniklaus. "Epigenetic Variation, Inheritance, and Selection in Plant Populations." In *Cold Spring Harbor Symposia on Quantitative Biology*, 97–104. Cold Spring Harbor, NY: Biological Laboratory Press, 2012.

Holeski, L. M., G. Jander, and A. A. Agrawal. "Transgenerational Defense Induction and Epigenetic Inheritance in Plants." *Trends in Ecology and Evolution* 27 (2012): 618–26.

Hollams, E. M., N. H. de Klerk, P. G. Hold, and P. D. Sly. "Persistent Effects of Maternal Smoking during Pregnancy on Lung Function and Asthma in Adolescents." *American Journal of Respiratory and Critical Care Medicine* 189 (2014): 401–7.

Horvath, S. "DNA Methylation Age of Human Tissues and Cell Types." *Genome Biology* 14 (2013): R115.

Horvath, S., W. Erhart, M. Brosch, O. Ammerpohl, W. von Schönfels, M. Ahrens, N. Heits, et al. "Obesity Accelerates Epigenetic Aging of Human Liver." *Proceedings of the National Academy of Sciences USA* 111 (2016): 15538–43.

Houri-Zeevi, L., and O. Rechavi. "A Matter of Time: Small RNAs Regulate the Duration of Epigenetic Inheritance." *Trends in Genetics* 33 (2017): 46–56.

Huang, P. H., and W. G. McBride. "Interaction of [Glutarimide-2-^{14}C]Thalidomide with Rat Embryonic DNA In Vivo." *Teratogenesis, Carcinogenesis, and Mutagenesis* 17 (1997): 1–5.

Hughes, K. A., and R. M. Reynolds. "Evolutionary and Mechanistic Theories of Aging." *Annual Review of Entomology* 50 (2005): 421–45.

Hutchison, C. A., III, R-Y. Chuang, V. N. Noskov, N. Assad-Garcia, T. J. Deerinck, M. H. Ellisman, J. Gill, et al. "Design and Synthesis of a Minimal Bacterial Genome." *Science* 351 (2016): aad6253.

Huxley, J. *Evolution: The Modern Synthesis*. New York: John Wiley and Sons, 1942 (reprinted 1964).

———. *Soviet Genetics and World Science*. London: Chatto and Windus, 1949.

Ishino, Y., H. Shinagawa, K. Makino, M. Amemura, and A. Nakata. "Nucleotide Sequence of the *iap* Gene, Responsible for Alkaline Phosphatase Isozyme Conversion in *Escherichia coli*, and Identification of the Gene Product." *Journal of Bacteriology* 169 (1987): 5429–33.

Itan, Y., B. L. Jones, C.J.E. Ingram, D. M. Swallow, and M. G. Thomas. "A Worldwide Correlation of Lactase Persistence Phenotype and Genotypes." *BMC Evolutionary Biology* 10 (2010): 36.

Jablonka, E. "Information: Its Interpretation, Its Inheritance, and Its Sharing." *Philosophy of Science* 69 (2002): 578–605.

Jablonka, E., and M. J. Lamb. *Epigenetic Inheritance and Evolution*. Oxford: Oxford University Press, 1995.

———. *Evolution in Four Dimensions*. Cambridge, MA: MIT Press, 2005.

Jablonka, E., and G. Raz. "Transgenerational Epigenetic Inheritance: Prevalence, Mechanisms, and Implications for the Study of Heredity and Evolution." *Quarterly Review of Biology* 84 (2009): 131–76.

Javurek, A. B., W. G. Spollen, A. M. Mann Ali, S. A. Johnson, D. B. Lubahn, N. J. Bivens, K. H. Bromert, et al. "Discovery of a Novel Seminal Fluid Microbiome and Influence of Estrogen Receptor Alpha Genetic Status." *Scientific Reports* 6 (2016): 23027.

———. "Corrigendum: Discovery of a Novel Seminal Fluid Microbiome and Influence of Estrogen Receptor Alpha Genetic Status." *Scientific Reports* 6 (2016): 25216.

Jennings, H. S. "Formation, Inheritance, and Variation of the Teeth in *Difflugia corona*: A Study of the Morphogenic Activities of Rhizopod Protoplasm." *Journal of Experimental Zoology* 77 (1937): 287–336.

Jeyaratnam, E., and S. Petrova. "Timeline: Key Events in the History of Thalidomide." *The Conversation* (2015). https://theconversation.com/timeline-key-events-in-the-history-of-thalidomide-50970.

Jiang, L., J. Zhang, J. J. Wang, L. Wang, L. Zhang, G. Li, X. Yang, et al. "Sperm, but Not Oocyte, DNA Methylome Is Inherited by Zebrafish Early Embryos." *Cell* 153 (2013): 773–84.

Jinek, M., K. Chylinski, I. Fonfara, M. Hauer, J. A. Doudna, and E. Charpentier. "Programmable Dual-RNA-Guided DNA Endonuclease in Adaptive Bacterial Immunity." *Science* 337 (2012): 816–21.

Johannes, F., and M. Colome-Tatche. "Quantitative Epigenetics through Epigenomic Perturbation of Isogenic Lines." *Genetics* 188 (2011): 215–27.

Johannes, F., E. Porcher, F. K. Teixeira, V. Saliba-Colombani, M. Simon, N. Agier, A. Bulski, et al. "Assessing the Impact of Transgenerational Epigenetic Variation on Complex Traits." *PLoS Genetics* 5 (2009): e1000530.

Johannsen, W. "The Genotype Conception of Heredity." *American Naturalist* 45 (1911): 129–59.

Johnstone, S. E., and S. B. Baylin. "Stress and the Epigenetic Landscape: A Link to the Pathobiology of Human Disease?" *Nature Reviews Genetics* 11 (2010): 806–12.

Jones, K. L., D. W. Smith, C. N. Ulleland, and A. P. Streissguth. "Pattern of Malformation in Offspring of Chronic Alcoholic Mothers." *The Lancet* 301 (1973): 1267–71.

Joseph, K. S., and M. S. Kramer. "Review of the Evidence on Fetal and Early Childhood Antecedents of Adult Chronic Disease." *Epidemiological Reviews* 18 (1996): 158–74.

Juliano, C. E., S. Z. Swartz, and G. M. Wessel. "A Conserved Germline Multipotency Program." *Development* 137 (2010): 4113–26.

Junnila, R. K., E. O. List, D. E. Berryman, J. W. Murrey, and J. J. Kopchick. "The GH/IGF-1 Axis in Ageing and Longevity." *Nature Reviews Endocrinology* 9 (2013): 366–76.

Kamimae-Lanning, A. N., S. M. Krasnow, N. A. Goloviznina, X. Zhu, Q. R. Roth-Carter, P. R. Levasseur, S. Jeng, et al. "Maternal High-Fat Diet and Obesity Compromise Fetal Hematopoiesis." *Molecular Metabolism* 4 (2014): 25–38.

Kanherkar, R. R., N. Bhatia-Dey, and A. B. Csoka. "Epigenetics across the Human Lifespan." *Frontiers in Cell and Developmental Biology* 2 (2014): 49.

Kasuga, T., and M. Gijzen. "Epigenetics and the Evolution of Virulence." *Trends in Microbiology* 21 (2013): 575–82.

Kawai, M. "Newly-Acquired Pre-Cultural Behaviour of the Natural Troop of Japanese Monkeys on Koshima Islet." *Primates* 6 (1965): 1–30.

Kern, S., M. Ackermann, S. C. Stearns, and T. J. Kawecki. "Decline in Offspring Viability as a Manifestation of Aging in *Drosophila melanogaster*." *Evolution* 55 (2001): 1822–31.

Kim, J. H., and A. R. Scialli. "Thalidomide: The Tragedy of Birth Defects and the Effective Treatment of Disease." *Toxicological Sciences* 122 (2011): 1–6.

Kirkpatrick, M., and R. Lande. "The Evolution of Maternal Characters." *Evolution* 43 (1989): 485–503.

Kirkwood, T. B., and M. R. Rose. "Evolution of Senescence: Late Survival Sacrificed for Reproduction." *Philosophical Transactions of the Royal Society of London B: Biological Sciences* 332 (1991): 15–24.

Kline, M. A., and R. Boyd. "Population Size Predicts Technological Complexity in Oceania." *Proceedings of the Royal Society B: Biological Sciences* 277 (2010): 2559–64.

Klironomos, F. D., J. Berg, and S. Collins. "How Epigenetic Mutations Can Affect Genetic Evolution: Model and Mechanism." *Bioessays* 35 (2013): 571–78.

Klosin, A., E. Casas, C. Hidalgo-Carcedo, T. Vavouri, and B. Lehner. "Transgenerational Transmission of Environmental Information in *C. elegans*." *Science* 356 (2017): 320–23.

Knightley, P., H. Evans, E. Potter, and M. Wallace. *Suffer the Children: The Story of Thalidomide*. London: Andre Deutsch, 1979.

Knopik, V. S., M. A. Maccani, S. Francazio, and J. E. McGeary. "The Epigenetics of Maternal Cigarette Smoking during Pregnancy and Effects on Child Development." *Developmental Psychopathology* 24 (2012): 1377–90.

Koestler, A. *The Sleepwalkers*. London: Arkana, 1959.

Kozak, G. M., M. L. Head, and J. W. Boughman. "Sexual Imprinting on Ecologically Divergent Traits Leads to Sexual Isolation in Sticklebacks." *Proceedings of the Royal Society B: Biological Sciences* 278 (2011): 2604–10.

Krutzen, M., J. Mann, M. R. Heithaus, R. C. Connor, L. Bejder, and W. B. Sherwin. "Cultural Transmission of Tool Use in Bottlenose Dolphins." *Proceedings of the National Academy of Sciences USA* 102 (2005): 8939–43.

Kuhn, T. S. *The Structure of Scientific Revolutions*. 2nd ed. Chicago: University of Chicago Press, 1970.

Kuijper, B., and R. A. Johnstone. "Maternal Effects and Parent-Offspring Conflict." *Evolution* (2017): DOI: 10.1111/evo.13403.

Lachlan, R. F., and M. R. Servedio. "Song Learning Accelerates Allopatric Speciation." *Evolution* 58 (2004): 2049–63.

Lachlan, R. F., M. N. Verzijden, C. S. Bernard, P-P. Jonker, B. Koese, S. Jaarsma, W. Spoor, P.J.B. Slater, and C. ten Cate. "The Progressive Loss of Syntactical Structure in Bird Song along an Island Colonization Chain." *Current Biology* 23 (2013): 1896–901.

Lachmann, M., and E. Jablonka. "The Inheritance of Phenotypes: An Adaptation to Fluctuating Environments." *Journal of Theoretical Biology* 181 (1996): 1–9.

Lacomble, S., S. Vaughan, C. Gadelha, M. K. Morphew, M. K. Shaw, J. R. McIntosh, and K. Gull. "Basal Body Movements Orchestrate Membrane Organelle Division and Cell Morphogenesis in *Trypanosoma brucei*." *Journal of Cell Science* 123 (2010): 2884–91.

Lafon-Placette, C., and C. Köhler. "Epigenetic Mechanisms of Postzygotic Reproductive Isolation in Plants." *Current Opinion in Plant Biology* 23 (2015): 39–44.

Lakatos, I. *The Methodology of Scientific Research Programmes*. Philosophical Papers 1. Cambridge: Cambridge University Press, 1978.

Laland, K. N., K. Sterelny, J. Odling-Smee, W. Hoppitt, and T. Uller. "Cause and Effect in Biology Revisited: Is Mayr's Proximate-Ultimate Distinction Still Useful?" *Science* 334 (2011): 1512–16.

Laland, K. N., T. Uller, M. W. Feldman, K. Sterelny, G. B. Müller, A. Moczek, E. Jablonka, and J. Odling-Smee. "The Extended Evolutionary Synthesis: Its Structure, Assumptions, and Predictions." *Proceedings of the Royal Society B: Biological Sciences* 282 (2015): 20151019.

Lamarck, J-B. *Philosophie Zoologique* (Zoological Philosophy). Translated by H. Elliot. London: Macmillan, 1809.

Laron, Z. "Laron Syndrome (Primary Growth Hormone Resistance or Insensitivity): The Personal Experience, 1958–2003." *Journal of Clinical Endocrinology and Metabolism* 89 (2004): 1031–44.

Lartigue, C., J. I. Glass, N. Alperovich, R. Pieper, P. P. Parmar, C. A. Hutchison 3rd., H. O. Smith, and J. C. Venter. "Genome Transplantation in Bacteria: Changing One Species to Another." *Science* 317 (2007): 632–38.

Latif, M., and N. S. Keenlyside. "El Niño/Southern Oscillation Response to Global Warming." *Proceedings of the National Academy of Sciences USA* 106 (2009): 20578–83.

LaTuga, M. S., A. Stuebe, and P. C. Seed. "A Review of the Source and Function of Microbiota in Breast Milk." *Seminars in Reproductive Medicine* 32 (2014): 68–73.

Laubichler, M. D., and E. H. Davidson. "Boveri's Long Experiment: Sea Urchin Merogones and the Establishment of the Role of Nuclear Chromosomes in Development." *Developmental Biology* 314 (2008): 1–11.

Laurance, J. "Experts Doubt Claims That Thalidomide Can Be Inherited." *The Independent*, August 11, 1997.

Lederberg, J. "Problems in Microbial Genetics." *Heredity* 2 (1948): 145–98.

Ledford, H. "CRISPR, the Disruptor." *Nature* 522 (2015): 20–24.

Leimar, O., and J. M. McNamara. "The Evolution of Transgenerational Integration of Information in Heterogeneous Environments." *American Naturalist* 185 (2015): E55–E69.

Lemoine, P., H. Harousseau, J. P. Borteyru, and J. C. Menuet. "Les enfants de parents alcooliques: Anomalies observées, a propos de 127 cas." *Quest-Médical* 21 (1968): 476–82.

Leslie, F. M. "Multigenerational Epigenetic Effects of Nicotine on Lung Function." *BMC Medicine* 11 (2013): 27.

Li, L., P. Zheng, and J. Dean. "Maternal Control of Early Mouse Development." *Development* 137 (2010): 859–70.

Liao, J., R. Karnik, H. Gu, M. J. Ziller, K. Clement, A. M. Tsankov, V. Akopian, et al. "Targeted Disruption of *DNMT1*, *DNMT3A*, and *DNMT3B* in Human Embryonic Stem Cells." *Nature Genetics* 47 (2015): 469–78.

Lieberman, D. E. *The Story of the Human Body: Evolution, Health, and Disease*. New York: Vintage, 2014.

Linnaeus, C. *Systema Naturae*. 10th ed. Vol. 1. Stockholm, Sweden: Holmiæ (Salvius), 1758.

Liu, S., K. Sun, T. Jiang, and J. Feng. "Natural Epigenetic Variation in Bats and Its Role in Evolution." *Journal of Experimental Biology* 218 (2015): 100–106.

Lockshon, D. "A Heritable Structural Alteration of the Yeast Mitochondrion." *Genetics* 161 (2002): 1425–35.

López-Beltrán, C. "The Medical Origins of Heredity." In *Heredity Produced: At the Crossroads of Biology, Politics, and Culture, 1500–1870*, edited by S. Müller-Wille and H-J. Rheinberger, 105–32. Cambridge, MA: MIT Press, 2007.

Lovelie, H., E. Immonen, E. Gustavsson, E. Kazancıoğu, and G. Arnqvist. "The Influence of Mitonuclear Genetic Variation on Personality in Seed Beetles." *Proceedings of the Royal Society B: Biological Sciences* 281 (2014): 20141039.

Luncz, L.V., R. Mundry, and C. Boesch. "Evidence for Cultural Differences between Neighboring Chimpanzee Communities." *Current Biology* 22 (2012): 922–26.

Lynch, M. "Mutation and Human Exceptionalism: Our Future Genetic Load." *Genetics* 202 (2016): 869–75.

Maddalo, D., E. Manchado, C. P. Concepcion, C. Bonetti, J. A. Vidigal, Y. C. Han, P. Ogrodowski, et al. "In Vivo Engineering of Oncogenic Chromosomal Rearrangements with the CRISPR/Cas9 System." *Nature* 516 (2014): 423–27.

Maegawa, S., Y. Lu, T. Tahara, J. T. Lee, J. Madzo, S. Liang, J. Jelinek, R. J. Coleman, and J-P. J. Issa. "Caloric Restriction Delays Age-Related Methylation Drift." *Nature Communications* 8 (2017): 53.

Maestripieri, D., and J. M. Mateo. *Maternal Effects in Mammals*. Chicago: University of Chicago Press, 2009.

Makino, H., A. Kushiro, E. Ishikawa, H. Kubota, A. Gawad, T. Sakai, K. Oishi, et al. "Mother-to-Infant Transmission of Intestinal Bifidobacterial Strains Has an Impact on the Early Development of Vaginally Delivered Infant's Microbiota." *PLoS One* 8 (2013): e78331.

Malagnac, F., and P. Silar. "Non-Mendelian Determinants of Morphology in Fungi." *Current Opinion in Microbiology* 6 (2003): 641–45.

Manikkam, M., R. Tracey, C. Guerrero-Bosagna, and M. K. Skinner. "Plastics Derived Endocrine Disruptors (BPA, DEHP, and DBP) Induce Epigenetic Transgenerational Inheritance of Obesity, Reproductive Disease and Sperm Epimutations." *PLoS One* 8 (2013): e55387.

Manolio, T. A., F. S. Collins, N. J. Cox, D. B. Goldstein, L. A. Hindorff, D. J. Hunter, M. I. McCarthy, et al. "Finding the Missing Heritability of Complex Diseases." *Nature* 461 (2009): 747–53.

Marlow, F. L. "Maternal Control of Development in Vertebrates." In *Colloquium Series on Developmental Biology*, edited by D. Kessler. San Rafael, CA: Morgan and Claypool Life Sciences, 2010.

Marshall, D. J., and T. Uller. "When Is a Maternal Effect Adaptive?" *Oikos* 116 (2007): 1957–63.

Mashoodh, R., B. Franks, J. P. Curley, and F. A. Champagne. "Paternal Social Enrichment Effects on Maternal Behavior and Offspring Growth." *Proceedings of the National Academy of Sciences USA* 109 (2012): 17232–38.

Matsuzawa, T., and W. C. McGraw. "Kinji Imanishi and 60 Years of Japanese Primatology." *Current Biology* 18 (2008): R587–R91.

Maynard Smith, J., and E. Szathmáry. *The Major Transitions in Evolution*. Oxford: W. H. Freeman, 1995.

Mayr, E. *The Growth of Biological Thought: Diversity, Evolution, and Inheritance*. Cambridge, MA: Belknap Press of Harvard University Press, 1982.

———. "Prologue: Some Thoughts on the History of the Evolutionary Synthesis." In *The Evolutionary Synthesis: Perspectives on the Unification of Biology*, edited by E. Mayr and W. B. Provine, 1–48. Cambridge, MA: Harvard University Press, 1998.

McBride, W. G., and P. A. Read. "Thalidomide May Be a Mutagen." *British Medical Journal* 308 (1998): 1635.

McDonald, J. I., H. Celik, L. E. Rois, G. Fishberger, T. Fowler, R. Rees, A. Kramer, et al. "Reprogrammable CRISPR/Cas9-Based System for Inducing Site-Specific DNA Methylation." *Biology Open* (2016): bio.019067.

McGhee, K. E., and A. M. Bell. "Paternal Care in a Fish: Epigenetics and Fitness Enhancing Effects on Offspring Anxiety." *Proceedings of the Royal Society B: Biological Sciences* 281 (2014): 20141146.

McGhee, K. E., L. M. Pintor, E. L Suhr, and A. M. Bell. "Maternal Exposure to Predation Risk Decreases Offspring Antipredator Behaviour and Survival in Threespined Stickleback." *Functional Ecology* 26 (2012): 932–40.

Mead, E. A., and D. K. Sarkar. "Fetal Alcohol Spectrum Disorders and Their Transmission through Genetic and Epigenetic Mechanisms." *Frontiers in Genetics* 5 (2014): Article 154.

Melentijevic, I., M. L. Toth, M. L. Arnold, R. J. Guasp, G. Harinath, K. C. Nguyen, D. Taub, et al. "*C. elegans* Neurons Jettison Protein Aggregates and Mitochondria under Neurotoxic Stress." *Nature* 542 (2017): 367–71.

Mesoudi, A. *Cultural Evolution: How Darwinian Theory Can Explain Human Culture and Synthesize the Social Sciences*. Chicago: University of Chicago Press, 2011.

Miller, G. A., M. S. Islam, T.D.W. Claridge, T. Dodgson, and S. J. Simpson. "Swarm Formation in the Desert Locust *Schistocerca gregaria*: Isolation and NMR Analysis of the Primary Maternal Gregarizing Agent." *Journal of Experimental Biology* 211 (2008): 370–76.

Miller, M. T., and K. Stromland. "Thalidomide: A Review, with a Focus on Ocular Findings and New Potential Uses." *Teratology* 60 (1999): 306–21.

Mojica, F. J., C. Díez-Villaseñor, J. García-Martínez, and E. Soria. "Intervening Sequences of Regularly Spaced Prokaryotic Repeats Derive from Foreign Genetic Elements." *Journal of Molecular Evolution* 60 (2005): 174–82.

Moran, N. A., and D. B. Sloan. "The Hologenome Concept: Helpful or Hollow?" *PloS Biology* 13 (2015): e1002311.

Moreira-Leite, F. F., T. Sherwin, L. Kohl, and K. Gull. "A Trypanosome Structure Involved in Transmitting Cytoplasmic Information during Cell Division." *Science* 294 (2001): 610–12.

Morgado, L., V. Preite, C. Oplaat, S. Anava, J. Ferreira de Carvalho, O. Rechavi, F. Johannes, and K. J. F. Verhoeven. "Small RNAs Reflect Grandparental Environments in Apomictic Dandelion." *Molecular Biology and Evolution* 34 (2017): 2035–40.

Morgan, H. D., F. Santos, K. Green, W. Dean, and W. Reik. "Epigenetic Reprogramming in Mammals." *Human Molecular Genetics* 14 (2005): R47–R58.

Morgan, H. D., H.G.E. Sutherland, D.I.K. Martin, and E. Whitelaw. "Epigenetic Inheritance at the *Agouti* Locus in the Mouse." *Nature Genetics* 23 (1999): 314–18.

Morgan, T. H. *The Physical Basis of Heredity*. Philadelphia: J. B. Lippincott, 1919.

———. *The Theory of the Gene*. New Haven, CT: Yale University Press, 1926.

Mostowy, R., J. Engelstadter, and M. Salathe. "Non-genetic Inheritance and the Patterns of Antagonistic Coevolution." *BMC Evolutionary Biology* 12 (2012): 93.

Mousseau, T. A., and H. Dingle. "Maternal Effects in Insect Life Histories." *Annual Review of Entomology* 36 (1991): 511–34.

Mousseau, T. A., and C. W. Fox. "The Adaptive Significance of Maternal Effects." *Trends in Ecology and Evolution* 13 (1998): 403–7.

———, eds. *Maternal Effects as Adaptations*. New York: Oxford University Press, 1998.

Moylan, S., K. Gustavson, S. Overland, E. B. Karevold, F. N. Jacka, J. A. Pasco, and M. Berk. "The Impact of Maternal Smoking during Pregnancy on Depressive and Anxiety Behaviors in Children: The Norwegian Mother and Child Cohort Study." *BMC Medicine* 13 (2015): 24.

Müller-Wille, S., and H-J. Rheinberger, eds. *A Cultural History of Heredity*. Chicago: University of Chicago Press, 2012.

———, eds. *Heredity Produced: At the Crossroads of Biology, Politics, and Culture, 1500–1870*. Cambridge, MA: MIT Press, 2007.

Nakagami, A., T. Negishi, K. Kawasaki, N. Imai, Y. Nishida, T. Ihara, Y. Kuroda, Y. Yoshikawa, and T. Koyama. "Alterations in Male Infant Behaviors towards Its Mother by Prenatal Exposure to Bisphenol-A in Cynomolgus Monkeys (*Macaca fascicularis*) during Early Suckling Period." *Psychoneuroendocrinology* 34 (2009): 1189–97.

Nanney, D. L. "Cortical Patterns in Cellular Morphogenesis." *Science* 160 (1968): 496–502.

NCD Risk Factor Collaboration. "A Century of Trends in Adult Human Height." *eLife* 5 (2016): e13410.

Nelson, M. L., A. Dinardo, J. Hochberg, and G. J. Armelagos. "Mass Spectroscopic Characterization of Tetracycline in the Skeletal Remains of an Ancient Population from Sudanese Nubia 350–550 CE." *American Journal of Physical Anthropology* 143 (2010): 151–54.

Nelson, V. R., S. H. Spiezio, and J. H. Nadeau. "Transgenerational Genetic Effects of the Paternal Y Chromosome on Daughters' Phenotypes." *Epigenomics* 2 (2010): 513–21.

Neuwald, M. F., M. Agranonik, A. K. Portella, A. Fleming, A. Wazana, M. Steiner, R. D. Livitan, M. J. Meaney, and P. P. Silveira. "Transgenerational Effects of Maternal Care Interact with Fetal Growth and Influence Attention Skills at 18 Months of Age." *Early Human Development* 90 (2014): 241–46.

Ng, S-F., R.C.Y. Lin, D. R. Laybutt, R. Barres, J. A. Owens, and M. J. Morris. "Chronic High-Fat Diet in Fathers Programs Β-Cell Dysfunction in Female Rat Offspring." *Nature* 467 (2010): 963–67.

Nieuwdorp, M., P. W. Gilijamse, N. Pai, and L. M. Kaplan. "Role of the Microbiome in Energy Regulation and Metabolism." *Gastroenterology* 146 (2014): 1525–33.

Nilsson, E. E., and M. K. Skinner. "Environmentally Induced Epigenetic Transgenerational Inheritance of Disease Susceptibility." *Translational Research* 1 (2015): 12–17.

Noble, D.W.A., R. W. Byrne, and M. J. Whiting. "Age-Dependent Social Learning in a Lizard." *Biology Letters* 10 (2014): 20140430.

Oberdoerffer, P., and D. A. Sinclair. "The Role of Nuclear Architecture in Genomic Instability and Ageing." *Nature Reviews Molecular Cell Biology* 8 (2007): 692–702.

O'Dea, R. E., D.W.A. Noble, S. L. Johnson, D. Hasselson, and S. Nakagawa. "The Role of Non-genetic Inheritance in Evolutionary Rescue: Epigenetic Buffering, Heritable Bet Hedging, and Epigenetic Traps." *Environmental Epigenetics* 2 (2016): 1–12.

Odling-Smee, F. J., K. N. Laland, and M. W. Feldman. *Niche Construction: The Neglected Process in Evolution*. Princeton, NJ: Princeton University Press, 2003.

Ost, A., A. Lempradl, E. Casas, M. Weigert, T. Tiko, M. Deniz, L. Pantano, et al. "Paternal Diet Defines Offspring Chromatin State and Intergenerational Obesity." *Cell* 159 (2014): 1352–64.

Ott, S. R., and S. M. Rogers. "Gregarious Desert Locusts Have Substantially Larger Brains with Altered Proportions Compared with the Solitarious Phase." *Proceedings of the Royal Society B: Biological Sciences* 277 (2010): 3087–96.

Painter, R., C. Osmond, P. Gluckman, M. Hanson, D. Phillips, and T. Roseboom. "Transgenerational Effects of Prenatal Exposure to the Dutch Famine on Neonatal Adiposity and Health in Later Life." *BJOG* 115 (2008): 1243–49.

Pál, C., and I. Miklós. "Epigenetic Inheritance, Genetic Assimilation, and Speciation." *Journal of Theoretical Biology* 200 (1999): 19–37.

Parker, G. A., R. R. Baker, and V. G. Smith. "The Origin and Evolution of Gamete Dimorphism and the Male-Female Phenomenon." *Journal of Theoretical Biology* 36 (1972): 529–53.

Pauly, P. J. "How Did the Effects of Alcohol on Reproduction Become Scientifically Uninteresting?" *Journal of the History of Biology* 29 (1996): 1–28.

Pembrey, M., R. Saffery, L. O. Bygren, and Network in Epigenetic Epidemiology. "Human Transgenerational Responses to Early-Life Experience: Potential Impact on Development, Health, and Biomedical Research." *Medical Genetics* 51 (2014): 563–72.

Pembrey, M. E., L. O. Bygren, G. Kaati, S. Edvinsson, K. Northstone, M. Sjostrom, J. Golding, and ALSPAC Study Team. "Sex-Specific, Male-Line Transgenerational Responses in Humans." *European Journal of Human Genetics* 14 (2006): 159–66.

Perry, J. C., L. K. Sirot, and S. Wigby. "The Seminal Symphony: How to Compose an Ejaculate." *Trends in Ecology and Evolution* 28 (2013): 414–22.

Peters, J. A., ed. *Classic Papers in Genetics*. Englewood Cliffs, NJ: Prentice-Hall, 1959.

Pfennig, D. W., and M. R. Servedio. "The Role of Transgenerational Epigenetic Inheritance in Diversification and Speciation." *Non-genetic Inheritance* 2012 (2012): 17–26.

Piotrowska, K., and M. Zernicka-Goetz. "Role for Sperm in Spatial Patterning of the Early Mouse Embryo." *Nature* 409 (2001): 517–21.

Pittet, F., M. Coignard, C. Houdelier, M-A. Richard-Yris, and S. Lumieau. "Effects of Maternal Experience on Fearfulness and Maternal Behaviour in a Precocial Bird." *Animal Behaviour* 85 (2013): 797–805.

Plaistow, S. J., C. Shirley, H. Collin, S. J. Cornell, and E. D. Harney. "Offspring Provisioning Explains Clone-Specific Maternal Age Effects on Life History and Life Span in the Water Flea, *Daphnia pulex*." *American Naturalist* 186 (2015): 376–89.

Popper, K. R. *Conjectures and Refutations: The Growth of Scientific Knowledge*. New York: Harper Torchbooks, 1968.

Potok, M. E., D. A. Nix, T. J. Parnell, and B. R. Cairns. "Reprogramming the Maternal Zebrafish Genome after Fertilization to Match the Paternal Methylation Pattern." *Cell* 153 (2013): 759–72.

Poulin, R., and F. Thomas. "Epigenetic Effects of Infection on the Phenotype of Host Offspring: Parasites Reaching across Host Generations." *Oikos* 117 (2008): 331–35.

Pourcel, C., G. Salvignol, and G. Vergnaud. "CRISPR Elements in *Yersinia pestis* Acquire New Repeats by Preferential Uptake of Bacteriophage DNA, and Provide Additional Tools for Evolutionary Studies." *Microbiology* 151 (2005): 653–63.

Prasad, N. G., S. Dey, A. Joshi, and T.N.C. Vidya. "Rethinking Inheritance, Yet Again: Inheritomes, Contextomes, and Dynamic Phenotypes." *Journal of Genetics* 94 (2015): 367–76.

Price, G. R. "The Nature of Selection." *Journal of Theoretical Biology* 175 (1995): 389–96.

———. "Science and the Supernatural." *Science* 122 (1955): 359–67.
———. "Selection and Covariance." *Nature* 227 (1970): 520–21.
———. "Where Is the Definitive Experiment?" *Science* 123 (1956): 17–18.
Proulx, S. R., and H. Teotónio. "What Kind of Maternal Effects Can Be Selected for in Fluctuating Environments?" *American Naturalist* 189 (2017): e118–37.
Qutob, D., B. P. Chapman, and M. Gijzen. "Transgenerational Gene Silencing Causes Gain of Virulence in a Plant Pathogen." *Nature Communications* 4 (2013): 1349.
Radzvilavicius, A. L. "Evolutionary Dynamics of Cytoplasmic Segregation and Fusion: Mitochondrial Mixing Facilitated the Evolution of Sex at the Origin of Eukaryotes." *Journal of Theoretical Biology* 404 (2016): 160–68.
Rassoulzadegan, M., V. Grandjean, P. Gounon, S. Vincent, I. Gillot, and F. Cuzin. "RNA-Mediated Non-Mendelian Inheritance of an Epigenetic Change in the Mouse." *Nature* 441 (2006): 469–74.
Reardon, S. "Welcome to the CRISPR Zoo." *Nature* 531 (2016): 160–63.
Rice, S. H. "The Place of Development in Mathematical Evolutionary Theory." *Journal of Experimental Zoology B: Molecular and Developmental Evolution* 318 (2012): 480–88.
Richards, C. L., C. Alonso, C. Becker, O. Bossdorf, E. Bucher, M. Colome-Tatche, W. Durka, et al. "Ecological Plant Epigenetics: Evidence from Model and Non-model Species, and the Way Forward." *Ecology Letters* 20 (2017): 1576–90.
Richards, C. L., O. Bossdorf, and M. Pigliucci. "What Role Does Heritable Epigenetic Variation Play in Phenotypic Evolution?" *BioScience* 60 (2010): 232–37.
Richards, C. L., O. Bossdorf, and K.J.F. Verhoeven. "Understanding Natural Epigenetic Variation." *New Phytologist* 187 (2010): 562–64.
Richards, E. J. "Inherited Epigenetic Variation—Revisiting Soft Inheritance." *Nature Reviews Genetics* 7 (2006): 395–401.
Richerson, P. J., and R. Boyd. "A Dual Inheritance Model of the Human Evolutionary Process I: Basic Postulates and a Simple Model." *Journal of Social and Biological Structures* 1 (1978): 127–54.
———. *Not by Genes Alone: How Culture Transforms Human Evolution*. Chicago: University of Chicago Press, 2005.
Riley, E. P., M. A. Infante, and K. R. Warren. "Fetal Alcohol Spectrum Disorders: An Overview." *Neuropsychological Review* 21 (2011): 73–80.
Robertson, M., and C. Richards. "Non-genetic Inheritance in Evolutionary Theory: The Importance of Plant Studies." *Non-genetic Inheritance* 2 (2015): 3–11.
Rodgers, A. B., C. P. Morgan, N. A. Leu, and T. L Bale. "Transgenerational Epigenetic Programming via Sperm microRNA Recapitulates Effects of Paternal Stress." *Proceedings of the National Academy of Sciences USA* 112 (2015): 13699–704.
Roll-Hansen, N. "Sources of Wilhelm Johannsen's Genotype Theory." *Journal of the History of Biology* 42 (2009): 457–93.
Rooij, S. R. de, H. Wouters, J. E. Yonker, R. C. Painter, and T. J. Roseboom. "Prenatal Undernutrition and Cognitive Function in Late Adulthood." *Proceedings of the National Academy of Sciences USA* 107 (2010): 16881–86.
Roseboom, T. J., J.H.P. van der Meulen, A.C.J. Ravelli, C. Osmond, D.J.P. Barker, and O. P. Bleker. "Effects of Prenatal Exposure to the Dutch Famine on Adult Disease in Later Life: An Overview." *Molecular and Cellular Endocrinology* 185 (2001): 93–98.
Rossen, N. G., J. K. MacDonald, E. M. de Vries, G. R. D'Haens, W. M. de Vos, E. G. Zoetendal, and C. Y. Ponsioen. "Fecal Microbiota Transplantation as Novel Therapy

in Gastroenterology: A Systematic Review." *World Journal of Gastroenterology* 21 (2015): 5359–71.

Rotem, E., A. Loinger, I. Ronin, I. Levin-Reisman, C. Gabay, N. Shoresh, O. Biham, and N. Q. Balaban. "Regulation of Phenotypic Variability by a Threshold-Based Mechanism Underlies Bacterial Persistence." *Proceedings of the National Academy of Sciences USA* 107 (2010): 12541–46.

Rozzi, F.V.R., and J. M. Bermudez de Castro. "Surprisingly Rapid Growth in Neanderthals." *Nature* 428 (2004): 936–39.

Sampson, P. D., A. P. Streissguth, F. L. Bookstein, R. E. Little, S. K. Clarren, P. Dahaene, J. W. Hanson, and J. M. Graham Jr. "Incidence of Fetal Alcohol Syndrome and Prevalence of Alcohol-Related Neurodevelopmental Disorder." *Teratology* 56 (1997): 317–26.

Santure, A. W., and H. G. Spencer. "Influence of Mom and Dad: Quantitative Genetic Models for Maternal Effects and Genomic Imprinting." *Genetics* 173 (2006): 2297–316.

Sapolsky, R. M. "Social Status and Health in Humans and Other Animals." *Annual Review of Anthropology* 33 (2004): 393–418.

Sapolsky, R. M., and L. J. Share. "A Pacific Culture among Wild Baboons: Its Emergence and Transmission." *PloS Biology* 2 (2004): e106.

Sapp, J. *Beyond the Gene: Cytoplasmic Inheritance and the Struggle for Authority in Genetics*. Oxford: Oxford University Press, 1987.

———. "Cytoplasmic Heretics." *Perspectives in Biology and Medicine* 41 (1988): 224–42.

———. *Genesis: The Evolution of Biology*. Oxford: Oxford University Press, 2003.

———. *Where the Truth Lies: Franz Moewus and the Origins of Molecular Biology*. New York: Cambridge University Press, 1990.

Sardet, C., F. Prodon, R. Dumollard, P. Chang, and J. Chenevert. "Structure and Function of the Egg Cortex from Oogenesis through Fertilization." *Developmental Biology* 241 (2002): 1–23.

Scheinfeld, A. "You and Heredity." In *A Treasury of Science*, edited by H. Shapley, S. Rapport, and H. Wright, 521–39. New York: Harper and Brothers, 1943.

Schmauss, C., Z. Lee-McDermott, and L. R. Medina. "Trans-Generational Effects of Early Life Stress: The Role of Maternal Behaviour." *Scientific Reports* 4 (2014): 4873.

Schmitz, R. J., M. D. Schultz, M. A. Urich, J. R. Nery, M. Pelizzola, O. Libiger, A. Alix, et al. "Patterns of Population Epigenomic Diversity." *Nature* 495 (2013): 193–98.

Schulz, L. C. "The Dutch Hunger Winter and the Developmental Origins of Health and Disease." *Proceedings of the National Academy of Sciences USA* 107 (2010): 16757–58.

Schuster, A., M. K. Skinner, and W. Yan. "Ancestral Vinclozolin Exposure Alters the Epigenetic Transgenerational Inheritance of Sperm Small Noncoding RNAs." *Environmental Epigenetics* 2 (2016).

Seymour, D. K., and C. Becker. "The Causes and Consequences of DNA Methylome Variation in Plants." *Current Opinion in Plant Biology* 36 (2017): 56–63.

Shah, D., Z. Zhang, A. B. Khodursky, N. Kaldalu, K. Kurg, and K. Lewis. "Persisters: A Distinct Physiological State of *E. coli*." *BMC Microbiology* 6 (2006): 53.

Sharma, A. "Transgenerational Epigenetic Inheritance: Focus on Soma to Germline Information Transfer." *Progress in Biophysics and Molecular Biology* 113 (2013): 439–46.

Sharon, G., D. Segal, J. M. Ringo, A. Hafetz, I. Zilber-Rosenberg, and E. Rosenberg. "Commensal Bacteria Play a Role in Mating Preference of *Drosophila melanogaster*." *Proceedings of the National Academy of Sciences USA* 107 (2013): 20051–56.

Sheriff, M. J., C. J. Krebs, and R. Boonstra. "The Sensitive Hare: Sublethal Effects of Predator Stress on Reproduction in Snowshoe Hares." *Journal of Animal Ecology* 78 (2009): 1249–58.

Shirokawa, Y., and M. Shimada. "Cytoplasmic Inheritance of Parent-Offspring Cell Structure in the Clonal Diatom *Cyclotella meneghiniana*." *Proceedings of the Royal Society B: Biological Sciences* 283 (2016): 20161632.

Shorter, J., and S. Lindquist. "Prions as Adaptive Conduits of Memory and Inheritance." *Nature Reviews Genetics* 6 (2005): 435–540.

Simões, B. F., F. L. Sampaio, C. Jared, M. M. Antoniazzi, E. R. Loew, J. K. Bowmaker, A. Rodriguez, et al. "Visual System Evolution and the Nature of the Ancestral Snake." *Journal of Evolutionary Biology* 28 (2015): 1309–20.

Simons, R. L., M. K. Lei, S.R.H. Beach, R. A. Philibert, C. E. Cutrona, F. X. Gibbons, and A. Barr. "Economic Hardship and Biological Weathering: The Epigenetics of Aging in a U.S. Sample of Black Women." *Social Science and Medicine* 150 (2016): 192–200.

Simpson, S. J., and G. A. Miller. "Maternal Effects on Phase Characteristics in the Desert Locust, *Schistocerca gregaria*: A Review of Current Understanding." *Journal of Insect Physiology* 53 (2007): 869–76.

Simpson, S. J., and D. Raubenheimer. *The Nature of Nutrition: A Unifying Framework from Animal Adaptation to Human Obesity*. Princeton, NJ: Princeton University Press, 2012.

Sinclair, D. A., and P. Oberdoerffer. "The Ageing Epigenome: Damaged beyond Repair?" *Ageing Research Reviews* 8 (2009): 189–98.

Sjöström, H., and R. Nilsson. *Thalidomide and the Power of the Drug Companies*. Harmondsworth, UK: Penguin, 1972.

Skinner, M. K., C. Guerrero-Bosagna, M. M. Haque, E. E. Nilsson, J.A.H. Koop, S. A. Knutie, and D. H. Clayton. "Epigenetics and the Evolution of Darwin's Finches." *Genome Biology and Evolution* 6 (2014): 1972–89.

Skinner, M. K. "Endocrine Disruptor Induction of Epigenetic Transgenerational Inheritance of Disease." *Molecular and Cellular Endocrinology* 398 (2014): 4–12.

———. "Environmental Epigenetics and a Unified Theory of the Molecular Aspects of Evolution: A Neo-Lamarckian Concept That Facilitates Neo-Darwinian Evolution." *Genome Biology and Evolution* 7 (2015): 1296–302.

———. "Environmental Stress and Epigenetic Transgenerational Inheritance." *BMC Medicine* 12 (2014): 153.

Slabbekoorn, H., and T. B. Smith. "Bird Song, Ecology, and Speciation." *Proceedings of the Royal Society of London B: Biological Sciences* 357 (2002): 493–503.

Smedley, S. R., and T. Eisener. "Sodium: A Male Moth's Gift to Its Offspring." *Proceedings of the National Academy of Sciences USA* 93 (1996): 809–13.

Smith, C. C., and S. D. Fretwell. "The Optimal Balance between Size and Number of Offspring." *American Naturalist* 108 (1974): 499–506.

Smith, G., and M. G. Ritchie. "How Might Epigenetics Contribute to Ecological Speciation?" *Current Zoology* 59 (2013): 686–96.

Smith, T. A., M. D. Martin, M. Nguyen, and T. C. Mendelson. "Epigenetic Divergence as a Potential First Step in Darter Speciation." *Molecular Ecology* 25 (2016): 1883–94.

Smith, T. M., P. Tafforeau, D. J. Reid, J. Pouech, V. Lazzari, J.P. Zermeno, D. Guatelli-Steinberg, et al. "Dental Evidence for Ontogenetic Differences between Modern Humans and Neanderthals." *Proceedings of the National Academy of Sciences USA* 107 (2010): 20923–28.

Smithells, D. "Does Thalidomide Cause Second Generation Birth Defects?" *Drug Safety* 19 (1998): 339–41.

Sniekers, S., S. Stringer, K. Watanabe, P. R. Jansen, J.R.I. Coleman, E. Krapohl, E. Taskesen, et al. "Genome-Wide Association Meta-Analysis of 78,308 Individuals Identifies New Loci and Genes Influencing Human Intelligence." *Nature Genetics* 49 (2017): 1107–12.

Sollars, V., X. Lu, L. Xiao, X. Wang, M. D. Garfinkel, and D. M. Ruden. "Evidence for an Epigenetic Mechanism by Which Hsp90 Acts as a Capacitor for Morphological Evolution." *Nature Genetics* 33 (2003): 70–74.

Sonneborn, T. M. "Does Preformed Cell Structure Play an Essential Role in Cell Heredity?" In *The Nature of Biological Diversity*, edited by J. M. Allen. New York: McGraw-Hill, 1963.

Soubry, A., C. Hoyo, R. L. Jirtle, and S. K. Murphy. "A Paternal Environmental Legacy: Evidence for Epigenetic Inheritance through the Male Germ Line." *Bioessays* 36 (April 2014): 359–71.

Soucy, S. M., J. Huang, and J. P. Gogarten. "Horizontal Gene Transfer: Building the Web of Life." *Nature Reviews Genetics* 16 (2015): 472–82.

Soyfer, V. N. *Lysenko and the Tragedy of Soviet Science*. New Brunswick, NJ: Rutgers University Press, 1994.

Stanner, S. A., and J. S. Yudkin. "Fetal Programming and the Leningrad Siege Study." *Twin Research* 4 (2001): 287–92.

Steele, E. J. *Somatic Selection and Adaptive Evolution: On the Inheritance of Acquired Characters*. Toronto: Williams and Wallace, 1979.

Steele, E. J., R. A. Lindley, and R. V. Blanden. *Lamarck's Signature: How Retrogenes Are Changing Darwin's Natural Selection Paradigm*. Frontiers of Science. Saint Leonards, NSW: Allen and Unwin, 1998.

Sterelny, K. *The Evolved Apprentice*. Cambridge, MA: MIT Press, 2012.

Sterelny, K., K. C. Smith, and M. Dickinson. "The Extended Replicator." *Biology and Philosophy* 11 (1996): 377–403.

Stockard, C. R, and G. N. Papanicolaou. "Further Studies on the Modification of the Germ-Cells in Mammals: The Effect of Alcohol on Treated Guinea-Pigs and Their Descendants." *Experimental Zoology* 26 (1918): 119–226.

Stoeckius, M., D. Grun, and N. Rajewsky. "Paternal RNA Contributions in the *Caenorhabditis elegans* Zygote." *EMBO Journal* 33 (2014): 1740–50.

Streissguth, A., H. Barr, J. Kogan, and F. Bookstein. "Primary and Secondary Disabilities in Fetal Alcohol Syndrome." In *The Challenge of Fetal Alcohol Syndrome*, edited by A. Streissguth and J. Kanter, 25–39. Seattle: University of Washington Press, 1997.

Stubbs, T. M., M. J. Bonder, A-K. Stark, F. Kruger, BI Ageing Clock Team, F. von Meyenn, O. Stegle, and W. Reik. "Multi-tissue DNA Methylation Age Predictor in Mouse." *Genome Biology* 18 (2017): 68.

Sultan, S. E. *Organism and Environment*. Oxford: Oxford University Press, 2015.

Sun, S. Y., C. Wang, Y. A. Yuan, and C. Y. He. "An Intracellular Membrane Junction Consisting of Flagellum Adhesion Glycoproteins Links Flagellum Biogenesis to Cell Morphogenesis in *Trypanosoma brucei*." *Journal of Cell Science* 126 (2012): 520–31.

Swallow, D. M. "Genetics of Lactase Persistence and Lactose Intolerance." *Annual Review of Genetics* 37 (2003): 197–219.
Swaminathan, N. "Not Milk? Neolithic Europeans Couldn't Stomach the Stuff." *Scientific American*, February 27, 2007.
Szathmáry, E. "The Evolution of Replicators." *Philosophical Transactions of the Royal Society of London B: Biological Sciences* 355 (2000): 1669–76.
Szyf, M. "Lamarck Revisited: Epigenetic Inheritance of Ancestral Odor Fear Conditioning." *Nature Neuroscience* 17 (2014): 2–4.
Tal, O., E. Kisdi, and E. Jablonka. "Epigenetic Contribution to Covariance between Relatives." *Genetics* 184 (2010): 1037–50.
Tanaka, S., and K. Maeno. "A Review of Maternal and Embryonic Control of Phase-Dependent Progeny Characteristics in the Desert Locust." *Journal of Insect Physiology* 56 (2010): 911–18.
Tharpa, A. P., M. V. Maffinia, P. A. Hunt, C. A. VandeVoort, C. Sonnenschein, and A. M. Soto. "Bisphenol A Alters the Development of the Rhesus Monkey Mammary Gland." *Proceedings of the National Academy of Sciences USA* 109 (2012): 8190–95.
Thompson, J. N., and J. J. Burdon. "Gene-for-Gene Coevolution between Plants and Parasites." *Nature* 360 (1992): 121–25.
Tolrian, R. "Predator-Induced Morphological Defences: Costs, Life History Shifts, and Maternal Effects in *Daphnia pulex*." *Ecology* 76 (1995): 1691–705.
Torres, R., H. Drummond, and A. Velando. "Parental Age and Lifespan Influence Offspring Recruitment: A Long-Term Study in a Seabird." *PLoS One* 6 (2011): e27245.
Uller, T. "Developmental Plasticity and the Evolution of Parental Effects." *Trends in Ecology and Evolution* 23 (2008): 432–38.
Uller, T., and I. Pen. "A Theoretical Model of the Evolution of Maternal Effects under Parent-Offspring Conflict." *Evolution* 65 (2011): 2075–84.
Uller, T., and H. Helanterä. "Niche Construction and Conceptual Change in Evolutionary Biology." *British Journal for the Philosophy of Science* (2017): DOI: 10.1093/bjps/axx050.
———. "Non-genetic Inheritance in Evolutionary Theory: A Primer." *Non-genetic Inheritance* 1 (2013): 27–32.
Uller, T., S. Nakagawa, and S. English. "Weak Evidence for Anticipatory Parental Effects in Plants and Animals." *Journal of Evolutionary Biology* 26 (2013): 2161–70.
Valtonen, T. M., K. Kangassalo, M. Polkki, and M. Rantala. "Transgenerational Effects of Parental Larval Diet on Offspring Development Time, Adult Body Size, and Pathogen Resistance in *Drosophila melanogaster*." *PLoS One* 7 (2012): e31611.
Van Leeuwen, E.J.C., K. A. Cronin, and D.B.M. Haun. "A Group-Specific Arbitrary Tradition in Chimpanzees (*Pan troglodytes*)." *Animal Cognition* 17 (2014): 1421–25.
Van Valen, L. "A New Evolutionary Law." *Evolutionary Theory* 1 (1973): 1–30.
Vargas, A. O. "Did Paul Kammerer Discover Epigenetic Inheritance? A Modern Look at the Controversial Midwife Toad Experiments." *Journal of Experimental Zoology B: Molecular and Developmental Evolution* 312 (2009): 667–78.
Vastenhouw, N. L., K. Brunschwig, K. L. Okihara, F. Müller, M. Tijsterman, and R.H.A. Plasterk. "Gene Expression: Long-Term Gene Silencing by RNAi." *Nature* 442 (2006): 882.

Vaughan, S., and H. R. Dawe. "Common Themes in Centriole and Centrosome Movements." *Trends in Cell Biology* 21 (2010): 57–66.

Veenendaal, M., R. Painter, S. de Rooij, P. Bossuyt, J. van der Post, P. Gluckman, M. Hanson, and T. Roseboom. "Transgenerational Effects of Prenatal Exposure to the 1944–45 Dutch Famine." *BJOG* 120 (2013): 548–54.

Vijendravarma, R. K., S. Narasimha, and T. J. Kawecki. "Effects of Parental Larval Diet on Egg Size and Offspring Traits in *Drosophila*." *Biology Letters* 6 (2010): 238–41.

Vilcinskas, A., K. Stoecker, H. Schmidtberg, R. Rohrich, and H. Vogel. "Invasive Harlequin Ladybird Carries Biological Weapons against Native Competitors." *Science* 340 (2013): 862–63.

Villa, P., and W. Roebroeks. "Neandertal Demise: An Archaeological Analysis of the Modern Human Superiority Complex." *PLoS One* 9 (2014): e96424.

Vojta, A., P. Dobrinić, V. Tadić, L. Bočkor, P. Korać, B. Julg, M. Klasić, and V. Zoldoš. "Repurposing the CRISPR-Cas9 System for Targeted DNA Methylation." *Nucleic Acids Research* 44 (2016): 5615–28.

Vojtech, L., S. Woo, S. Hughes, C. Levy, L. Ballweber, R. P. Sauteraud, J. Strobl, et al. "Exosomes in Human Semen Carry a Distinctive Repertoire of Small Non-coding RNAs with Potential Regulatory Functions." *Nucleic Acids Research* 42 (2014): 7290–304.

Waal, F. B. M. de. *Are We Smart Enough to Know How Smart Animals Are?* New York: Norton, 2016.

Wade, N. "Scientists Seek Moratorium on Edits to Human Genome That Could Be Inherited." *New York Times*, December 3, 2015.

Wagner, K. D., N. Wagner, H. Ghanbarian, V. Grandjean, P. Gounon, F. Cuzin, and M. Rassoulzadegan. "RNA Induction and Inheritance of Epigenetic Cardiac Hypertrophy in the Mouse." *Developmental Cell* 14 (2008): 962–69.

Wang, T., B. Tsui, J. F. Kreisberg, N. A. Robertson, A. M. Gross, M. Ku Yu, H. Carter, H. M. Brown-Borg, P. D. Adams, and T. Ideker. "Epigenetic Aging Signatures in Mice Livers Are Slowed by Dwarfism, Calorie Restriction, and Rapamycin Treatment." *Genome Biology* 18 (2017): 57.

Wang, Y., H. Liu, and Z. Sun. "Lamarck Rises from His Grave: Parental Environment-Induced Epigenetic Inheritance in Model Organisms and Humans." *Biological Reviews of the Cambridge Philosophical Society* 92 (2017): 2084–111.

Wang, Z., Z. Wang, J. Wang, Y. Sui, J. Zhang, D. Liao, and R. Wu. "A Quantitative Genetic and Epigenetic Model of Complex Traits." *BMC Bioinformatics* 13 (2012): 274.

Warkany, J. "Why I Doubted That Thalidomide Was the Cause of the Epidemic of Limb Defects of 1959 to 1961." *Teratology* 38 (1988): 217–19.

Warner, R. H., and H. L. Rosett. "The Effects of Drinking on Offspring: A Historical Survey of the American and British Literature." *Journal of Studies on Alcohol* 36 (1975): 1395–420.

Waterland, R. A., R. Kellermayer, E. Laritsky, P. Rayco-Solon, R. A. Harris, M. Travisano, W. Zhang, et al. "Season of Conception in Rural Gambia Affects DNA Methylation at Putative Human Metastable Epialleles." *PLoS Genetics* 6 (2010): e1001252.

Watson, J.E.M., D. F. Shanahan, M. Di Marco, J. Allan, W. F. Laurance, E. W. Sanderson, B. Mackey, and O. Venter. "Catastrophic Declines in Wilderness Areas Undermine Global Environment Targets." *Current Biology* 26 (2016): 2929–34.

Watson, R., and E. Szathmáry. "How Can Evolution Learn?" *Trends in Ecology and Evolution* 31 (2016): 147–57.

Weismann, A. *Essays upon Heredity and Kindred Biological Problems*. Translated by E. B. Poulton, S. Schonland, and A. E. Shipley. Oxford: Clarendon Press, 1889.

———. *The Evolution Theory*. Translated by J. A. Thomson and M. R. Thomson. Vol. 2 London: Edward Arnold, 1904.

———. *The Germ-Plasm: A Theory of Heredity*. Translated by W. Newton Parker and H. Ronnfeldt. New York: Charles Scribner's Sons, 1893.

Weismann, G. "The Midwife Toad and Alma Mahler: Epigenetics or a Matter of Deception?" *FASEB* 24 (2010): 2591–95.

Werren, J. H., L. Baldo, and M. E. Clark. "*Wolbachia*: Master Manipulators of Invertebrate Biology." *Nature Reviews Microbiology* 6 (2008): 741–51.

West-Eberhard, M. J. "Dancing with DNA and Flirting with the Ghost of Lamarck." *Biology and Philosophy* 22 (2007): 439–51.

———. *Developmental Plasticity and Evolution*. Oxford: Oxford University Press, 2003.

Weyrich, L. S., S. Duchene, J. Soubrier, L. Arriola, B. Llamas, J. Breen, A. G. Morris, K. W. Alt, D. Caramelli, V. Dresely, M. Farrell, A. G. Farrer, M. Francken, N. Gully, W. Haak, K. Hardy, K. Harvati, P. Held, E. C. Holmes, J. Kaidonis, C. Lalueza-Fox, M. de la Rasilla, A. Rosas, P. Semal, A. Soltysiak, G. Townsend, D. Usai, J. Wahl, D. H. Huson, K. Dobney, and A. Cooper. "Neanderthal Behaviour, Diet, and Disease Inferred from Ancient DNA in Dental Calculus." *Nature* 544 (2017): 357–61.

White, J., P. Mirleau, E. Danchin, H. Mulard, S. A. Hatch, P. Heeb, and R. H. Wagner. "Sexually Transmitted Bacteria Affect Female Cloacal Assemblages in a Wild Bird." *Ecology Letters* 13 (2010): 1515–24.

Whiten, A., J. Goodall, W. C. McGrew, T. Nishida, V. Reynolds, Y. Sugiyama, C.E.G. Tutin, R. W. Wrangham, and C. Boesch. "Charting Cultural Variation in Chimpanzees." *Behaviour* 138 (2001): 1481–516.

Wilson, T. *Distilled Spirituous Liquors the Bane of the Nation: Being Some Considerations Humbly Offer'd to the Legislature, Part II*. 2nd ed. London: J. Roberts, 1736.

Wolf, J. B., E. D. Brodie III, J. M. Cheverud, A. J. Moore, and M. J. Wade. "Evolutionary Consequences of Indirect Genetic Effects." *Trends in Ecology and Evolution* 13 (1998): 64–69.

Wolf, J. B., and M. J. Wade. "What Are Maternal Effects (and What Are They Not)?" *Philosophical Transactions of the Royal Society of London B: Biological Sciences* 364 (2009): 1107–15.

Wolff, G. L. "Influence of Maternal Phenotype on Metabolic Differentiation of *Agouti* Locus in the Mouse." *Genetics* 88 (1978): 529–39.

Wright, S. "The Roles of Mutation, Inbreeding, Crossbreeding, and Selection in Evolution." In *Proceedings of the Sixth International Congress of Genetics*, edited by D. F. Jones, 356–66. Austin, TX: Genetics Society of America, 1932.

Yan, M., Y. Want, Y. Hu, Y. Feng, C. Dai, J. Wu, D. Wu, Z. Fang, and Q. Zhai. "A High-Throughput Quantitative Approach Reveals More Small RNA Modifications in Mouse Liver and Their Correlation with Diabetes." *Analytical Chemistry* 85 (2013): 12173–81.

Yehuda, R., N. P. Daskalakis, L. M. Bierer, H. N. Bader, T. Klengel, F. Holsboer, and E. B. Binder. "Holocaust Exposure Induced Intergenerational Effects on *FKBP5* Methylation." *Biological Psychiatry* 80 (2015): 372–80.

Yehuda, R., M. H. Teicher, J. R. Seckl, R. A. Grossman, A. Morris, and L. M. Bierer. "Parental PTSD as a Vulnerability Factor for Low Cortisol Trait in Offspring of Holocaust Survivors." *Archives of General Psychiatry* 64 (2007): 1040–48.

Youngson, N. A., and E. Whitelaw. "Transgenerational Epigenetic Effects." *Annual Review of Genomics and Human Genetics* 9 (2008): 233–57.

Zalasiewicz, J., M. Williams, A. Smith, T. L. Barry, A. L. Coe, P. R. Brown, P. Brenchley, et al. "Are We Now Living in the Anthropocene?" *GSA Today* 18 (2008): DOI: 10.1130/gsat018 02A.1.

Zamenhof, S., E. van Marthens, and L. Grauel. "DNA (Cell Number) in Neonatal Brain: Second Generation (F2) Alteration by Maternal (F0) Dietary Protein Restriction." *Science* 172 (1971): 850–51.

Zamenhof, S., E. van Marthens, and F. L. Margolis. "DNA (Cell Number) and Protein in Neonatal Brain: Alteration by Maternal Dietary Protein Restriction." *Science* 160 (1968): 322–23.

Zannas, A. S., J. Arloth, T. Carrillo-Roa, S. Iurato, S. Röh, K. J. Ressler, C. B. Nemeroff, et al. "Lifetime Stress Accelerates Epigenetic Aging in an Urban, African American Cohort: Relevance of Glucocorticoid Signaling." *Genome Biology* 16 (2015): 266.

Zimmer, C. "What Is a Species?" *Scientific American*, June 2008.

Zirkle, C. "The Early History of the Idea of the Inheritance of Acquired Characters and of Pangenesis." *Transactions of the American Philosophical Society* 35 (1946): 91–151.

Ziv-Gal, A., W. Wang, C. Zhou, and J. A. Flaws. "The Effects of In Utero Bisphenol A Exposure on Reproductive Capacity in Several Generations of Mice." *Toxicology and Applied Pharmacology* 284 (2015): 354–62.

Zoghbi, H. Y., and A. L. Beaudet. "Epigenetics and Human Disease." *Cold Spring Harbor Perspectives in Biology* 8 (2016): a019497.

INDEX